EXPLANATIONS

OF

THE PRINCIPLES OF ARITHMETIC,

ON

A NEW PLAN.

BY CORNELL MOREY,

TEACHER OF MATHEMATICS IN MACEDON ACADEMY.

ROCHESTER:

JOHN GREVES, PRINTER, 22 STATE-STREET.

1850.

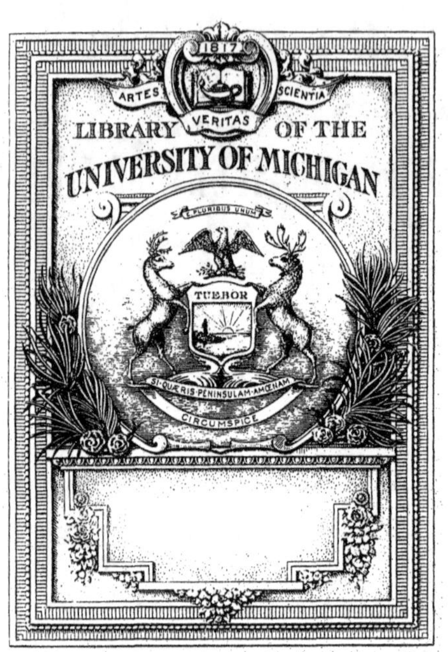

RECEIVED IN EXCHANGE
FROM
United States Library
of Congress

EXPLANATIONS

OF

THE PRINCIPLES OF ARITHMETIC,

ON

A NEW PLAN.

BY CORNELL MOREY,

TEACHER OF MATHEMATICS IN MACEDON ACADEMY.

ROCHESTER:

JOHN GREVES, PRINTER, 22 STATE-STREET.

1850.

Entered according to Act of Congress, in the year 1850, by
CORNELL MOREY,
in the Clerk's Office of the District Court of the Northern District
of New-York.

The following pages have been prepared amid the labors of the school-room, and with a view to lessen those labors on the part of the teacher, as well as to aid the pupil in obtaining a thorough and practical knowledge of the principles of ARITHMETIC.

That something of the kind is much needed, to accomplish these ends, every dutiful teacher at once admits.

That no improvement can be made upon this production, the author does not claim: but should the system of teaching which it anticipates be approved, the author will soon present to the public, a work upon this branch of Mathematics, which by itself will serve as a text-book, and which perhaps will be more acceptable than the present volume.

The author wishes not to *force* upon the public an additional work upon this subject, but he only asks for a careful examination of this production, and of the system of teaching the science which it proposes, leaving its introduction into use to the discretion of teachers.

An expression of the conclusion arrived at by those who do thus investigate, will be kindly received by C. MOREY, at Macedon Centre, Wayne Co., N. Y.

AUGUST 1, 1850.

ARITHMETIC.

ARITHMETIC comprises both a science and an art. A science, so far as it involves the relations and properties of numbers, as they exist in nature, and an art, so far as these relations and properties are exhibited and applied by means of characters.

NOTATION AND NUMERATION.

Notation consists in writing numbers by figures.

Numeration consists in reading the values of figures when written.

To represent the absence of every thing, I write that character, (0) called naught.
To represent that I have one unit of any kind, I write that character, (1) called one.
..................two units..............................(2) .... two.
..................three..................................(3) .... three.
..................four...................................(4) ... four.
..................five...................................(5) ... five.
..................six....................................(6) ... six.
..................seven..................................(7) .... seven.
..................eight..................................(8) ... eight.
..................nine...................................(9) ... nine.

Now if I represent every different number by a different character; as we have an *infinite number* of numbers, I should necessarily have an *infinite number* of *characters*. Hence it becomes necessary to combine these characters so that they may be made to represent every different number.

Therefore if I wish to represent that I have ten units of any kind, I will do it by writing the 1 one place towards the left, and filling the place it formerly occupied with a naught: thus the value of ten is given to 1 written in the second place from the right - - - - 10

Ten and one are eleven - - - 11
Ten and two are twelve - - - 12
Ten and three are thirteen - - 13
Ten and five are fifteen - - - 15
Ten and nine are nineteen - - 19

As twenty is composed of two tens I will represent it by writing 2 in ten's place - - 20

Twenty and two are twenty-two - - 22

Twenty and five are twenty-five - - 25

Twenty and nine are twenty-nine - - 29

As thirty is composed of three tens, I will represent it by writing 3 in ten's place - - 30

To represent ninety, write 9 in ten's place - 90

Ninety and nine are ninety-nine - - 99

As one hundred is composed of ten tens, I will represent it by writing 1 one place farther towards the left, or in the third place from the right - - - - - 100

We may observe that in accordance with the system now introduced, by removing a figure one place towards the left, it is made to represent a value ten times as great.

The next figure at the right is called *Units* 2

The next is ten times units, called *Tens* 2

The next is ten times tens, called *Hundreds* 2

The next is ten times hundreds, called *Thousands* 2

The next is ten times thousands, called *Tens of Thousands* 2

The next is ten times tens of thousands, called *Hundreds of Thousands* 2

The next is ten times hundreds of thousands, called *Millions* 2

The next is ten times millions, called *Tens of Millions* 2

The next is ten times tens of millions, called *Hundreds of Millions* 2

The next is ten times hundreds of millions, called *Billions* 2

The next is ten times billions, called *Tens of Billions* 2

The next is ten times tens of billions, called *Hundreds of Billions* 2

The next is ten times hundreds of billions, called *Trillions* 2

The next is ten times trillions, or *Tens of Trillions* 2
&c., &c., &c.

Now as we do not read farther than Hundreds, in any

denomination, and as to represent Hundreds requires three figures, numbers become divided into periods of three figures each.

The first period at the right is called Units, from the Latin word *unus*, meaning one.

The second and third periods have *no appropriate* names.

The fourth period is called Billions. The prefix Bi. means two, and the number of the period may be found by adding two to the meaning of the prefix of the name.

The fifth period is called Trillions. The prefix Tri. means three. The number of *this* period may be found by adding two to the meaning of the prefix of the name.

The sixth period is called Quadrillions. The prefix Quad. means four. The number of *this* period may be found by adding two to the meaning of the prefix of the name.

And in *general*, for all periods at the left of the third, having the *name* given to find the *number*, add 2 to the meaning of the prefix of the name. And having the number of the period given, to find the name, take 2 from the number of the period, aud we have the meaning of the prefix of the name.

9th Sept.	8th Sex.	7th Quin.	6th Quad.	5th Tri.	4th Bi.	3d Mill.	2d Thou.	1st Units.
222	222	222	222	222	222	222	222	222

Suppose that I am required to arrange in accordance with this system these numbers,

Quadrillions	40	In this example the prefix Sex. has the greatest value ; hence the Sextillion's period will be at the left hand. And Sex. meaning six, the period will be the eighth.
Sextillions -	364	
Thousands -	.23	
Billions - -	489	
Units -	376	

The next prefix in value is Quad., which meaning four, the number of the period will be six, and as each period must occupy three places, I must write a naught at the left of the 40 to fill the vacant place. There being no Quintillions given, the seventh period must be filled with naughts.

There being no Trillions given, the *fifth* period must also be filled with naughts.

The next prefix in value is Bi., which meaning two, the 489 must be written in the fourth period.

There being no Millions given, the third period must be filled with naughts. The 23 must be written in the second period, and a naught be placed at the left of it that the period may occupy three places. The 376 being units, must be written in the first period.

Therefore, we have for the arrangement,

8th	7th	6th	5th	4th	3d	2d	1st
364	000	040	000	489	000	023	376

The nine figures that have value are called significant figures.

They are also sometimes called the nine digits.

The value a figure represents, when standing alone, is called its simple value.

The increased value which a figure represents, when others are written at the right of it, is called the local value.

ADDITION.

Addition consists in uniting several numbers into one.

I wish to add several numbers.

3588 As I cannot add units of one denomination to
217 units of a different denomination, I have written
4523 the numbers so that units of the *same* denomination
5466 shall fall under each other. As the values of
13794 figures increase from the right hand towards the left, I will commence at the right hand to add.

Six units, and three, and seven, and eight units are twenty-four units—equal to four units and two tens. I will write the 4 under the column of units, and add the 2 to the column of tens, that I may have all the tens collected together. Two tens and six, and two, and one, and eight tens are nineteen tens,—equal to nine tens and one hundred. I will write the 9 under the column of tens, and add the 1 to the column of hundreds, that I may have all the hundreds collected together. There are in all seventeen hundreds,—equal to seven hundreds and one thousand. I will write the 7 under the column of hundreds, and add the 1 to the column of thousands that I may have all the thousands collected together. There are in all thirteen thousands, which I will write down. The result obtained is called the amount or sum.

Hence to add numbers :—

Write them so that units of the same denomination shall fall under each other. Commence at the right hand and add each column separately. Write down the right hand figure of each amount, and add the left hand figure or figures to the next column. Thus proceed to the last column, where write down the whole amount.

SUBTRACTION.

Subtraction consists in finding the difference between two numbers.

From	7854	The greater number is called the *Minuend.*
I wish to take	3281	
	4573	The less number the *Subtrahend.*

The difference between them is called the *Remainder.*

As we cannot take units of one denomination from units of a different denomination, I have written the numbers so that units of the same denomination shall fall under each other. As the values of figures increase from the right hand towards the left, I will commence at the right hand to subtract. One unit taken from four units leaves three units. Eight tens from five tens I cannot take. I will add ten tens to the five tens, making fifteen tens. Eight tens from fifteen tens leave seven tens. Now as I have added ten tens or one hundred to the Minuend, the Remainder will be too large by one hundred, hence I must take away from the Minuend one hundred more than I at first intended to; that is, I must take away three hundreds instead of two hundreds. Three hundreds taken from eight hundreds leave five hundreds. Three thousands taken from seven thousands leave four thousands.

Hence to take one number from another:

Write the less number under the greater, so that units of the same denomination shall fall under each other. Commence at the right hand, and take each figure in the lower line from the one above it. If the value of a figure in the lower line is greater than the one above it, add ten to the upper figure and then subtract: then add one to the next left hand figure of the Subtrahend, and thus proceed.

It will appear evident, by examination, that any number may be added to the figure in the Minuend, that will render it of greater value than the figure in the Subtrahend. But to avoid fractions the number ten is used.

MULTIPLICATION.

3459 4 13836	I wish to obtain four times 3459.

As the values of figures increase from the right hand towards the left, I will commence at the right hand. Four times nine units are thirty-six units, equal to six units and three tens. I will write down the six under the units, reserving the three tens. Four times five tens are twenty tens and the three tens reserved being added, give twenty-three tens, equal to three tens and two hundreds. I will write the 3 under the tens, reserving the two hundreds. Four times four hundreds are sixteen hundreds, and the two hundreds reserved being added, give eighteen hundreds, equal to eight hundreds and one thousand. I will write the 8 under the hundreds, reserving the one thousand. Four times three thousands are twelve thousands, and the one thousand reserved being added, gives thirteen thousand, which I will write down. Hence the result is 13836.

This operation is called MULTIPLICATION. Hence Multiplication is taking one number as many times as there are units in another.

The number to be multiplied is called the *Multiplicand.*

The number by which we multiply is called the *Multiplier*, and the result obtained is called the *Product.*

Both Multiplier and Multiplicand are called *Factors.*

When the Multiplier does not exceed twelve;—Write the Multiplier under the right of the Multiplicand. Multiply each figure of the Multiplicand by the Multiplier, commencing at the right hand. Set down the right hand figure of each partial product, and add the left hand figure or figures to the product of the next figure of the Multiplicand.

When the Multiplier exceeds Twelve.

I wish to multiply four thousand two hundred and fifty-six by three hundred and twenty-five.

```
   4256
    325
  -----
  21280
  8512
 12768
-------
1383200
```

First I will take five times the Multiplicand, which gives this line (21280.) I will take two times the Multiplicand which gives this line (8512,)—but as the Multiplier is tens, I must write the Product one place towards the left. I will next take three times the Multiplicand, which gives this line (12768,)—but as the Multiplier is hundreds, I must write the product in hundreds' place. Now this line (12768) is three hundred times the Multiplicand, this one (8512) is twenty times, and this one (21280) is five times the Multiplicand. Hence if I add them all together, I shall have three hundred and twenty-five times the Multiplicand.

Therefore when the Multiplier exceeds twelve,—

Write the Multiplier under the Multiplicand. Multiply by each figure of the Multiplier separately, placing the right hand figure of each partial product directly under its Multiplier. Then the sum of the partial products will be the entire product.

When the Multiplier is a Composite Number.

I wish to multiply 4567 by 24.

4567 8 36536 3 109608	First I will multiply it by 8. Now as the multiplier is only one third part as large as it should be, the product is only one third of what it should be; hence, to make it what it should be, I must multiply it by three.

Such a number as 24, which is composed of two or more factors, is called a *Composite number*, and the *factors* of which it is composed are called the *Component parts*.

Hence when the Multiplier is a Composite number;—We may first multiply by one of those component parts, and then that product by another, and thus proceed until all the factors have been used. The last product will be the product required.

When there are Cyphers at the right of the Multiplier.

I wish to multiply 3467 by 2000.

3467 2000 6934000	If I multiply by the 2; as the multiplier will have been only one thousandth part as large as it should be, the product will be only

one thousandth part as large as *it* should be,—hence to make it what it should be, multiply by one thousand, which is evidently done by annexing three naughts. Hence multiply by the significant figures of the multiplier, and annex to the product as many naughts as are at the right of the multiplier.

It is evident that when there are cyphers at the right hand of *both* the factors, they may all be omittted until we shall have multiplied the significant figures, and then annex to the product, so obtained, as many naughts as there are at the right hand of both the factors.

DIVISION.

4) 3496 } Suppose that I wish to divide 3496 Apples
874 } among four individuals, and to find the number that I can give to each. Suppose that I have them arranged in heaps, three of which contain one thousand each, four heaps with one hundred in a heap, nine heaps with ten in a heap, and six single apples.

Not having as many heaps of one thousand in a heap as there are individuals, it is evident that I cannot give them one *such heap* a piece, hence I will change them into heaps of the next smaller size. One of the heaps with one thousand in a heap, is equivalent to ten heaps with one hundred in a heap, and three are equal to three times ten or thirty, and four heaps added give thirty four heaps with one hundred in a heap. Now, if I go around to the four individuals and give them one heap a piece, it will take four of them, and if I pass around eight times, it will take eight times four, or thirty-two heaps, and I shall have two heaps left. But as there are not enough heaps left to go around again, I will change the remainder to heaps of the *next* smaller size. One of the heaps with one hundred in a heap, is equivalent to ten heaps with ten in a heap, and two such heaps will be equal to two times ten or twenty, and nine heaps added give twenty-nine heaps with ten in a heap. Now, if I go around to the four individuals and give them one heap a piece, it will take four of them, and if I pass around seven times, it will take seven times four, or *twenty-eight* heaps, and I shall have one heap left, which contains ten single apples, and the six apples added give sixteen single apples. Now, if I give them one of these a piece, it will take four, and if I give them four apiece it will take *four times four*, and there will be none remaining.

This operation is called DIVISION. Hence Division consists in finding the number of times one number is contained in another.

The number to be divided is called the *Dividend*;

The number that we divide by is called the *Divisor* ; and the result obtained is called the *Quotient.*

I observe that I first found the number of times the Divisor is contained in as many of the left hand figures of the Dividend as are *necessary* to contain it, and I had a remainder of two. Then in changing to heaps of the next smaller size, and adding in those next smaller, I obtained 29, the same that I should have obtained by prefixing the remainder to the next figure of the Dividend. Hence the rule.

FOR SHORT DIVISION, OR

When the Divisor does not exceed Twelve.

Write the Divisor at the left hand of the dividend, and find the number of times that it is contained in as many of the left hand figures of the dividend as are necessary to contain it, and prefix the remainder (if any) to the next figure of the dividend, then divide, and thus proceed until all the figures of the dividend shall have been used.

When the Divisor exceeds Twelve.

```
24)6864(286
   48
   ---
   206
   192
   ---
    144
    144
    ---
```

Suppose that I wish to divide 6864 Apples among 24 individuals, and to find the number that I may give to each. Suppose that I have them arranged in heaps, six of which contain one thousand each, eight heaps with one hundred in a heap, six heaps with ten in a heap, and four single apples.

Not having as many heaps of one thousand in a heap, as there are individuals, it is evident that I cannot give them one *such heap* a piece, hence I will change them into heaps of the next smaller size. One of the heaps with one thousand in a heap is equivalent to ten heaps with one hundred in a heap, and six such will be equal to six times ten or sixty, and eight heaps added give sixty-eight heaps, with one hundred in a heap.

Now, if I go around to the twenty-four individuals, and give them one heap a-piece, it will take twenty-four of them, and if I pass around twice, it will take twice twenty-four or forty-eight heaps, which being taken from the first number, sixty-eight, leaves twenty heaps with one hundred in a heap, remaining.

But as there are not enough heaps left to go around again, I will change the remainder to heaps of the next smaller size. One of the heaps with one hundred in a heap is equivalent to ten heaps with ten in a heap, and twenty such are equal to twenty times ten, or two hundred, and six heaps added give two hundred and six heaps with ten in a heap.

Now if I go around to the twenty-four individuals, and give them one heap a piece, it will take twenty-four of them, and if I pass around eight times it will take eight times twenty-four or one hundred and ninety-two heaps, which being taken from the two hundred and six, leaves fourteen heaps with ten in a heap remaining. As there are not enough heaps left to go around again, I will change the remainder to heaps of the next smaller size or to single apples. In one of these heaps there are ten single apples and in fourteen of them there are fourteen times ten or one hundred and forty, and four added give one hundred and forty-four single apples. Now, if I give them six of these a-piece, it will take one hundred and forty-four, and there will be no remainder.

HENCE FOR LONG DIVISION, OR,

When the Divisor Exceeds Twelve.

Write the Divisor at the left of the dividend, and find the number of times it is contained in as many of the left hand figures of the dividend as are necessary to contain it: place the result at the right of the dividend, as the left hand figure of the quotient, multiply the divisor by this quotient figure, and subtract the product from that part of the dividend used : to the remainder annex the next figure of the dividend, then divide again, multiply and subtract as before, and thus proceed until all the figures of the dividend have been used.

When the Divisor is a Composite number.

8)8640 ⎫
6)1080 ⎬ I wish to divide 8640 by 48.
180 ⎭

First, I will divide by 8. Now, as the divisor has been only one-sixth part as large as it should be, the *quotient* is *six times* as large as it should be, hence, to make it what it should be I will divide *this* quotient by 6.

Hence, when the Divisor is a composite number—we may divide first by one of the component parts, and that quotient by another, continuing the operation until all the factors have been used.

It is evident from the system of Notation, that to divide by 10 or 100, &c., we have only to strike off from the right hand of the Dividend as many figures as there are naughts at the right hand of the divisor. Thus, dividing 768 by 10 gives for a quotient 76, and a remainder 8. And 8642 divided by 100 gives 86 for a quotient, and a remainder 42.

When there are Naughts at the right of the Divisor.

2ø) 345⁄7 ⎫ Suppose I have 3457 to be divided by
172+17 ⎭ 20. The Divisor is composed of two

factors, 2 and 10. To divide by 10 strike off the 7. Then divide the remaining part of the Dividend by 2, I have 1 for a remainder, which is evidently of an order of units next above those of the 7. And as the remainder obtained by the second division will always be of the order of tens, compared with the remainder obtained by the first division, we have the following Rule :

When there are Naughts at the right of the Divisor.

Strike off the naughts, and strike off as many figures at the right hand of the Dividend, then divide the remaining part of the Dividend by the significant figures of the Divisor. If a Remainder occur, annex to it the figures cut off from the Dividend, and we have the true Remainder.

After several Divisors have been used, to find the Remainder in units of the same denomination of the original Dividend.

Division		Remainders
4)35678		2 = 2
5)8919+2		4+4 = 16
9)1783+4	EXAMPLE.	5+4+1 = 20
7)198+1		9+5+4×2=360
28+2		Entire rem. 398

I have divided the number 35678 by the factors of 1260, and have found several Remainders. I wish now to find the *true Remainder* in units of the same denomination of the *original Dividend.* The remainder 2 is a part of the original dividend, that has not been divided, hence it will form a part of the true remainder. The second remainder 4 is a part of the second dividend,—but as the second dividend is found by dividing the first dividend by 4, this remainder must evidently be multiplied by 4. The third remainder 1 is a part of the third dividend,—but as the third dividend is found by dividing by 4 and 5, it is evident that this remainder must be multiplied by 4 and 5 to change *it* to units of the same denomination of the original dividend. The fourth remainder 2 is a part of the fourth dividend, but as the fourth dividend is found by dividing by 4, 5 and 9, it is evident that this remainder must be multiplied by 4, 5, and 9, to change it to units of the same denomination of the original dividend. Notice that the last remainder is multiplied by all the divisors before it, except its own, by the 9, 5, and 4, and the same is true of *each* remainder. Hence the rule.

Multiply each remainder by all the divisions before it, except the one immediately preceding, add the several products, thus obtained, together, and their sum will be the true remainder in units of the same denomination of the original dividend.

REDUCTION.

Reduction consists in changing the number of units of one denomination to that of another, *without changing their value.*

I wish to find what number of pence there are in - - - - - - -

£	s.	d.
12	8	7
20		
248		
12		
2983		

As there are twenty shillings in one pound, there will be twenty times as many shillings as pounds. Hence I will multiply the number of pounds by 20, to find the number of shillings. Twenty times twelve give two hundred and forty, and adding the eight shillings, we have two hundred and forty-eight shillings. Now as twelve pence make one shilling, there will be twelve times as many pence as shillings. Hence I will now multiply the number of shillings by 12, to find the number of pence. Twelve times two hundred and forty-eight give 2976, and adding the seven pence, we obtain 2983 pence for the result.

This operation is called REDUCTION. When, as in this example, units of higher denominations are to be reduced to units of lower denominations, it is called *Reduction Descending*, to perform which, commence with the highest denomination and multiply by the number of units that it takes of the next lower denomination to make one of that, adding in the units of the next lower denomination, if there be any, and thus proceed until the quantity is reduced to the denomination required.

When units of lower denominations are being reduced to units of higher denominations, the operation is called *Reduction Ascending*, and as it is the reverse of Reduction Descending, we shall, evidently, have the following rule:

Divide the given number by the number of units of that lower denomination required to make one of the next higher, and thus continue to the denomination required. If remainders occur, they will be of the same denomination as the dividend from which they are obtained.

ADDITION OF COMPOUND NUMBERS.

Compound Numbers are such as represent units of different denominations, and do not increase or decrease in a tenfold proportion.

I wish to find the sum of

£	*s.*	*d.*	*far.*
5	19	7	2
6	17	8	1
9	12	3	3
22	9	7	2

As the values of these numbers increase towards the left, I will commence at the right hand to add. Three and one and two farthings give six farthings. Four farthings make one penny—hence, in six farthings there are as many pence as 4 is contained times in 6. It is contained once, and there are two farthings remaining, which I will write under the column of farthings, and I will add the one penny to the column of pence, that I may have all the pence collected together.

One and three and eight and seven pence give nineteen pence. Twelve pence make one shilling—hence, in nineteen pence there are as many shillings as 12 is contained times in 19. It is contained once, and there are seven pence remaining, which I will write under the column of pence, and I will add the one shilling to the column of shillings, that I may have all the shillings collected together.

One and twelve and seventeen and nineteen shillings give forty-nine shillings. Twenty shillings make one pound—hence, in forty-nine shillings there are as many pounds as 20 is contained times in 49. It is contained two times, and there are nine shillings remaining, which I will write under the column of shillings, and I will add the two pounds to the column of pounds, that I may have all the pounds collected together.

Two and nine and six and five pounds give twenty-two pounds. Therefore the whole amount is £22 9*s.* 7*d.* 2*far.*

Hence, to Add Compound Numbers :

Write them so that units of the same denomination shall fall under each other. Commence at the right hand, and add the units of each denomination separately, dividing each amount obtained by the number of units of that denomination required to make one of the next higher. Write down the remainder, (if any,) and add the quotient to the next higher denomination, with which proceed as before.

SUBTRACTION OF COMPOUND NUMBERS.

	£	*s.*	*d.*	*far.*
From	9	7	5	3
I wish to take	3	9	8	2
	5	17	9	1

Commencing at the right hand : two farthings from three farthings leave one farthing. Eight pence from five pence I cannot take. I will add twelve pence to the five pence, making seventeen pence. Eight pence from seventeen pence leave nine pence. But, now, as I have added twelve pence or one shilling to the minuend, the remainder will be too large by one shilling ; hence, I must take away from the minuend one shilling more than I at first intended to. That is, instead of taking away nine shillings, I must take away ten shillings. But I cannot take ten shillings from seven shillings. I will add twenty shillings to the seven shillings, making twenty-seven shillings. Ten from twenty-seven leaves seventeen shillings. But, now, as I have added twenty shillings or one pound to the minuend, the remainder will be too large by one pound ; hence, I must take away from the minuend one pound more than I at first intended to. That is, instead of taking away three pounds, I must take away four pounds. Four pounds from nine pounds leave five pounds.

Hence, to Subtract Compound Numbers:

Write the less number under the greater, with units of the same denomination under each other. Commence at the right hand and take each number in the lower line from the one above it. When the number of units in any denomination of the subtrahend exceeds the number in the same denomination of the minuend, add to that denomination of the minuend as many units as make one of the next higher denomination, and then subtract; after which, add one to the units of the next higher denomination of the subtrahend, and thus proceed.

MULTIPLICATION OF COMPOUND NUMBERS.

	£	s.	d.	
I wish to multiply	8	13	5	by 7
			7	
	60	13	11	

As the values of the numbers increase from the right towards the left, I will commence at the right hand. Seven times five pence give thirty-five pence. Twelve pence make one shilling; hence, in thirty-five pence there are as many shillings as 12 is contained times in 35. It is contained twice, and there are eleven pence remaining, which I will write under the pence, reserving the two shillings.

Seven times thirteen shillings give ninety-one shillings, and the two shillings reserved being added give ninety-three shillings. Twenty shillings make one pound; hence, in ninety-three shillings there are as many pounds as 20 is contained times in 93. It is contained four times, and there are thirteen shillings remaining, which I will write under the shillings, reserving the four pounds.

Seven times eight pounds give fifty-six pounds, and the four pounds reserved being added, give sixty pounds, which I will write down.

Hence, to Multiply Compound Numbers:

Commence at the right hand and multiply the units of

the lowest denomination by the multiplier. Divide the product thus obtained by the number of units of that denomination required to make one of the next higher. Write down the remainder, and add the quotient to the product of the next higher denomination. Divide as before, and thus proceed to the highest denomination, where write down the whole product.

When the Multiplier is a Composite Number.

It is evident that we may use its factors, in the same manner as in simple numbers.

When the Multiplier is large, and is not a Composite Number.

Multiply the price of a unit by 10, to obtain the cost of ten units; then multiply this product by 10, to obtain the cost of one hundred units; then multiply the cost of 100 units by the number of hundreds, the cost of 10 units by the number of tens, and the cost of one unit by the number of units; and the sum of these several products will be the product required.

DIVISION OF COMPOUND NUMBERS.

	£	s.	d.
4)	38	9	8
	9	12	5

Suppose that I wish to divide £38 9*s*. 8*d*. among four individuals, and to find the sum that may be given to each. If I go around to the four individuals and give them one pound apiece, it will evidently take four of them, and if I pass around nine times, it will take nine times four pounds, or thirty-six pounds, and I shall have two pounds remaining. But as there are not enough pounds left to go around again, I will change this remainder to shillings. There are twenty shillings in one pound, hence there will be twenty times as many shillings as there are pounds.

Twenty times two give forty, and the nine shillings added give forty-nine shillings. Now, if I go around to the four individuals and give them one shilling apiece, it will take four of them, and if I pass around twelve times it will take twelve times four, or forty-eight shillings, and I shall have one shilling left, which is equal to twelve pence, and the eight pence added give twenty pence, of which if I give them five pence apiece there will be none remaining. Hence, I can give to each of the four individuals £9 12*s.* 5*d.*

Hence, for the Division of Compound Numbers:

Write the divisor at the left hand of the dividend, and find the number of times it is contained in the units of the highest denomination; write the quotient underneath, and multiply the remainder, (if any,) by the number of units of the next lower denomination required to make one of this higher, and add to this product the given units of the next lower denomination, (if any,) then divide by the given divisor, and thus proceed to the lowest denomination.

It is evident that when the divisor is a composite number, the component parts may be used, as in simple numbers.

When the divisor is large, and is not a composite number, represent the work after the manner of Long Division, instead of retaining the whole in the mind—thus:

```
        £    s.  d.   £  s.  d.
257 ) 561   2   4 ( 2   3   8
      514
      ---
       47
       20
      ---
      942
      771
      ---
      171
       12
      ----
     2056
     2056
```

An *odd number* is a number that cannot be divided by 2 without a remainder; as 3, 7, 13.

An *even number* is a number that can be divided by 2 without a remainder; as 6, 10, 14.

All numbers are either odd or even.

The sum or difference of any two odd numbers, or of any two even numbers, is an even number.

But the sum or difference of an odd and even number is an odd number.

The sum of any number of even numbers is an even number; hence, if an even number be multiplied by any number whatever, the product will be even.

The sum of an even number of odd numbers is an even number, but

The sum of an odd number of odd numbers is an odd number; hence,

The product of two or more odd numbers is an odd number.

A product is even if one of the factors which compose it is even.

An odd number cannot be divided by an even number without a remainder.

A number is said to divide another when it is contained in that number without a remainder.

A *prime* number is a number that can be divided only by itself and a unit; as 7, 11, 19.

A *factor* of a number is one of the numbers which being multiplied together will produce the given number. Thus 3 and 7, are the factors of 21.

A *prime* factor is a factor of a number which can be divided only by itself and a unit.

A *composite* number is a number that is composed of two or more factors.

A *prime* number cannot be resolved into factors.

All numbers are either prime or composite.

A composite number is resolved into its *prime factors*, when each of the factors into which it is resolved is a prime number.

The number 15 is composed of the factors 3 and 5 : if we divide it by 3, we obtain the 5 for a quotient; that is, we reject in the dividend a factor the size of the divisor.

Again ; the number 28 is composed of the factors 4 and 7,—if we divide it by 4 we obtain the 7,—that is, we reject in the dividend a factor the size of the divisor, and the remaining factor is the quotient.

And in general, Division may be said to consist in striking out of the dividend a factor the size of the divisor.

And Multiplication may be said to consist in combining with the multiplicand such factors as occur in the multiplier.

Hence,—it will appear evident, that if we multiply by a number, and then divide the product by the *same* number, we will not change the value.

Again ; if we are to multiply and then to divide the product by a number containing a factor common to both multiplier and divisor, that factor may be omitted.

Again ; as division consists in striking out of the dividend a factor the size of the divisor, it is evident that a dividend cannot contain a divisor, unless there is a factor of that divisor in the dividend. For instance, 15 will contain 3 as a divisor, for 3 is one of the factors of 15. But 21 will not contain 5 as a divisor, for 5 is not a factor in 21.

Again ; it will appear evident, that if a divisor is a composite number, the dividend which contains it must be made up in *part*, at least, of the prime factors of that divisor—that is, in whatever dividend a divisor is contained, the prime factors of that divisor are contained in the same dividend.

Again ; it will be observed that the quotient is always the factor which is left after striking out of the dividend a factor the size of the divisor. Hence, if we multiply the dividend by any factor, we increase the value of the quotient by the same factor—and if we divide the dividend, we diminish the value of the quotient.

If we multiply the divisor, we diminish the value of the quotient—for in performing the division, we reject more factors in the dividend, and hence have less left for the quotient.

And if we diminish the divisor by any factor, we increase the quotient by the same factor.

If we multiply or divide *both* divisor and dividend by the same number, we shall not affect the value of the quotient.

From the preceding remarks, it is plain that the dividend is the product of the divisor and quotient—that the divisor may be found by dividing the dividend by the quotient—that if we multiply together the factors given, we obtain a product,—and that if we divide the product of any number of factors by the product of a part of those factors, we shall obtain the product of the remaining factors for a quotient—and if we divide the product of any number of factors by the product of all but one of them, we shall obtain that one for a quotient.

THE LEAST COMMON MULTIPLE.

3)	18	12	9	5
3)	6	4	3	5
2)	2	4	1	5
	1	2	1	5

$3 \times 3 \times 2 \times 2 \times 5 = 180$ least common multiple.

A multiple of a number is a number that may be divided by that number without a remainder. Thus, 12 is a multiple of 4. 18 is a multiple of 9, &c.

A common multiple of two or more numbers is such a number as may be divided by each of them, *separately*, without a remainder. Thus 24 is a common multiple of 4, 3, and 6.

The *least* common multiple of several numbers is the *least* number that may be divided by each of them without a remainder. Thus 24 is the least common multiple of 6, 4 and 8.

I wish to find the least common multiple of the numbers 18, 12, 9 and 5.

If I find a number that may be divided by each of these numbers without a remainder, it must be composed of the prime factors of each of these numbers.

If I find the *least number* that may be divided by each of them without a remainder, *it* must be made up of the prime factors of these numbers taken but *once each*. Hence, if any two or more of these numbers contain a common prime factor, that factor must be used once and *but* once.

Therefore for the purpose of eliminating such superfluous factors as may exist in these numbers, I will divide by any prime factor that is common to two or more of them—reserving that factor once in the divisor, and throwing it out of each of the numbers in which it occurs.

I notice that a factor 3 is common to three of the numbers, hence I will divide by 3. I notice that there is *another* factor 3 common to two of the numbers ; hence I will divide *again* by 3. There is yet a factor 2, common to two of the numbers, hence I will now divide by 2. There are now no two numbers that contain a common factor. Hence, I conclude, as I have saved each of the prime factors of these numbers once and *but once*, that the product of the divisors, quotients and undivided numbers, will be the least common multiple.

Hence to find the least common multiple of two or more numbers :

Write the numbers in a line, and divide by any prime number that is an exact divisor of two or more of them ; write the quotients and undivided numbers as a second horizontal line, with which proceed as before—continue the division until no two of the numbers contain a common factor. Then the product of the divisors, quotients and undivided numbers, will be the least common multiple required.

THE GREATEST COMMON DIVISOR.

```
252     252)616(3                  | 16 | 4
616         504                 4· | 20 | 5
            ---                    | —— | —
            112)252(2              | 36 | 9
                224
                ---
                28)112(4
                   112      616=252×2+112.
```

A number that is contained in two or more numbers an exact number of times, is called their *Common Divisor*. And the greatest number that is contained in two or more numbers an exact number of times is called their *Greatest Common Divisor*.

I wish to find the greatest common divisor of two numbers, as 252 and 616. I will first illustrate a principle on which this demonstration depends. If any number (as 36) be separated into two parts, (as 16 and 20,) and each of the parts and the entire number be divided by the same divisor, the sum of the partial quotients thus obtained will be equal to the entire quotient. And if the entire quotient and one of the partial quotients are whole numbers, the other partial quotient must be, for no proper fraction added to a whole number will give a whole number. And if both the partial quotients are whole numbers, the entire quotient must be, for no two whole numbers when added give a fraction.

I know that no number larger than 252 can be the greatest common divisor of 252 and 616, for no number greater than 252 is contained in 252. Hence, if 252 is an exact divisor of 616, it is the greatest common divisor of the two numbers.

I find by dividing, that there is a remainder, (112.) hence, 252 is not that greatest common divisor. Now if I can show that the greatest common divisor of the less number and remainder after division, is the greatest common divisor of the greater and less, I will then proceed to find the greatest common divisor of the less number and remainder, and it will be the greatest common divisor required.

First, I will show that the greatest common divisor of the greater and less is a common divisor of the less and remainder.

The greater is always made up of some multiple of the less, plus the remainder. It is a divisor of the greater by hypothesis ; it is also a divisor of the less by hypothesis, and it is evidently a divisor of any multiple of the less : hence, I have only to show that it is a divisor of the remainder. It must be a divisor of the remainder, for the sum of the partial quotients is equal to the entire quotient, and the entire quotient and one of the partial quotients are whole numbers : hence, the other partial quotient must be a whole number, for no proper fraction added to a whole number will give a whole number.

Next, I will show that the greatest common divisor of the less and remainder, is a common divisor of the greater and less.

It is a divisor of the less by hypothesis—it is evidently a divisor of any multiple of the less—and it is a divisor of the remainder by hypothesis—hence, it must be a divisor of the greater ; for, the sum of the partial quotients is equal to the entire quotient, and both the partial quotients being whole numbers, the entire quotient must be a whole number, for no two whole numbers added will give a fraction.

Now I claim, that the greatest common divisor of the less and remainder, is the *greatest* common divisor of the greater and less. For suppose the greater and less have a common divisor, *greater* than the greatest common divisor of the less and remainder,—it could not be contained in the less and remainder—but I have shown that it must be contained in them. Hence, the greater and less cannot have a common divisor *greater* than the greatest common divisor of the less and remainder.

Again—suppose the greatest common divisor of the greater and less to be *less* than the greatest common divisor of the less and remainder—then the greater and less could not contain the greatest common divisor of the less and remainder,—but I have shown that they must contain it.

Hence, I conclude that the greatest common divisor of the greater and less cannot be either *greater* or *less* than the greatest common divisor of the less and remainder—therefore it must be the *same number*.

Hence, I will now proceed to find the greatest common divisor of the less and remainder, and it will be the greatest common divisor sought.

I know that no number greater than 112 can be that greatest common divisor, for no number greater than 112 is contained in 112; therefore if it is an exact divisor of 252, it is the greatest common divisor sought. I find there is a remainder, (28.)

Now, considering this 112 as the less and 28 as the remainder, I will proceed to find their greatest common divisor. Dividing 112 by 28 I have no remainder; hence, 28 is the greatest common divisor of 28 and 112; therefore, from what has been shown, it must be the greatest common divisor of 112 and 252; hence, 28 is also the greatest common divisor of 252 and 616.

Therefore, to find the greatest common divisor of two numbers:

Divide the greater number by the less, and the last divisor by the last remainder, continuing the operation until nothing remains; and the last divisor will be the greatest common divisor.

To find the greatest common divisor of more than two numbers:

First find the greatest common divisor of two of them, and then of that common divisor and another of the numbers; thus continuing until all the numbers have been used.

VULGAR FRACTIONS.

Introduction to Fractions.

$$8 \qquad 7 \qquad 5 \qquad 3$$

$$\tfrac{3}{8} \qquad \tfrac{2}{7} \qquad \tfrac{4}{5} \qquad \tfrac{1}{3}$$

$$\tfrac{5}{7} \text{ and } \tfrac{3}{5}\,;\ \tfrac{6}{6} \text{ and } \tfrac{8}{3}\,;\ 3\tfrac{1}{7}\,;\ \tfrac{3}{4} \text{ of } \tfrac{7}{8}\,;\ \frac{\frac{6}{11}}{\frac{9}{13}},\ \frac{\frac{3}{5}}{7} \text{ and } \frac{3}{\frac{4}{7}}$$

$$\tfrac{4}{8},\ \tfrac{3}{12},\ \tfrac{4}{20},\ \Big|\ \tfrac{12}{3}\quad 3)\tfrac{12}{4}\ \Big|\ 12)\tfrac{1728}{3456}\,;\ 12)\tfrac{144}{288}\,;\ 6)\tfrac{12}{24}\,;\ 2)\tfrac{2}{4}\,;\ \tfrac{1}{2}.$$

If I wish to represent that I have a unit of any kind divided into eight equal parts, I will do it by writing the figure 8. If I wish to represent that I have a unit divided into seven equal parts, I will do it by writing the figure 7. If I wish to represent that I have a unit divided into five equal parts, I will do it by writing the figure 5. If I wish to represent that I have a unit divided into three equal parts, I will do it by writing the figure 3.

If of these eight equal parts I wish to represent that I take three, I will do it by writing 3 above the 8 with a line drawn between them. If of the 7 equal parts I wish to represent that I take two, I will do it by writing 2 above the 7, with a line between them. If of the five equal parts I wish to represent that I take four, I will do it by writing 4 above the 5, with a line between them. If of the three equal parts I wish to represent that I take one, I will do it by writing 1 above the 3 with a line between them.

The number below the line in each case, shows the number of parts into which the unit is divided. And as these parts take their *name* from the number into which the unit is divided, the number below the line becomes the *namer* of the parts—hence it is called the *Denominator.*

The number above the line, shows the number of the parts that is taken, hence it becomes a *numberer*, and is therefore called the *Numerator.*

The whole expression represents a part of a unit, and hence it is called a *Fraction.*

Both Numerator and Denominator are called *terms* of the fraction.

When the numerator is less than the denominator, the expression represents a part of a unit—it is then called a *proper fraction.* As $\frac{5}{7}$ and $\frac{3}{5}$, (read five-sevenths and three-fifths.)

When the numerator is equal to, or greater than the denominator, the expression denotes something more than a unit, hence it is then called an *improper fraction.* As $\frac{6}{6}$ and $\frac{8}{3}$, (read six-sixths and eight-thirds.)

When we have a fraction joined to a whole number, the expression is called a *mixed number.* As $3\frac{5}{7}$, (read three and five-sevenths.)

A fraction of a fraction is called a *compound fraction.* As $\frac{3}{4}$ of $\frac{7}{8}$.

A fraction having a fraction for either its numerator or denominator, or both, is called a *complex fraction.* As $\frac{\frac{3}{5}}{7}$, $\frac{3}{\frac{4}{7}}$, $\frac{\frac{6}{11}}{\frac{9}{13}}$, (read three-fifths divided by seven, three divided by four-sevenths, and six-elevenths divided by nine-thirteenths.)

If I have a unit divided into eight equal parts, and I wish to take one-half that unit, it is evident that I must take one-half of the number of parts into which the unit is divided. That is, the numerator, which *shows* the number of parts I take, must be one-half of the denominator, which shows the number of parts into which the unit is divided.

If I have a unit divided into twelve equal parts, and I wish to take one-fourth of that unit, it is evident that I must take one-fourth of the number of parts into which the unit is divided; that is, the numerator, which *shows* the number of parts I take, must be *one-fourth* of the denominator.

If I have a unit divided into twenty equal parts, and I wish to take one-fifth of that unit, it is evident that I must take one-fifth of the number of parts into which the unit is divided; that is, the numerator, which *shows* the number of parts I take, must be *one-fifth* of the denominator.

And I conclude that, in general, *the numerator must be*

such a part of the denominator as the value of the fraction is intended to represent.

If I have units divided into four equal parts each, and I wish to take enough of those parts to make three whole ones, it is evident that I must take three times as many parts as compose one unit—that is, the numerator must be three times the denominator—hence, if the numerator be divided by the denominator, I shall obtain the value of the fraction, (3,) for a quotient. And in *general*, the value of a fraction is the quotient arising from dividing the numerator by the denominator.

In case of division, if we diminish both divisor and dividend by the same factor, we do not change the value of the quotient—and as to find the value of a fraction we perform division, using the numerator for a dividend and the denominator for a divisor; it is evident that we may divide both numerator and denominator of a fraction, by the same number, without changing the value.

This operation is called reducing fractions to their lowest terms. Hence, to reduce a fraction to its lowest terms:

Divide both numerator and denominator by any number that is an exact divisor of them both, and the resulting fraction in the same manner, thus continuing until no number, greater than a unit, will divide them both. Or we may find the greatest common divisor of both numerator and denominator, and then reduce it to its lowest terms by one division.

To change a Whole Number to an equivalent Fraction.

5 to 8ths. I wish to change a whole number to an equiv-
8 alent fraction, having a specified denominator; for
— instance, 5 to 8ths.
40 $\frac{40}{8}$ There are eight eighths in one whole one; that is, there are eight times as many eighths as there are whole ones; hence, in five whole ones there are eight *times five* eighths, or *forty* eighths.

Hence, to change a whole number to an equivalent fraction having a specified denominator:

Multiply the whole number by the specified denominator, and under that product write the denominator.

To change an Improper Fraction to a Whole or a Mixed Number.

$\frac{23}{7}$ $7)\underline{23}$ I wish to change an improper fraction to $3\frac{2}{7}$ an equivalent whole or mixed number, for instance $\frac{23}{7}$ to a whole or mixed number. There are seven-sevenths in one whole one, and in twenty-three sevenths there are as many whole ones as seven is contained times in twenty-three. Three and two-seventh times.

Hence, to change an improper fraction to an equivalent whole or mixed number:

Divide the numerator by the denominator.

To change a Mixed Number to an equivalent Fraction.

$4\frac{7}{8}$ to 8ths.	I wish to change a mixed number to an
4	equivalent fraction; for instance $4\frac{7}{8}$ to 8ths.
8	There are eight eighths in one whole one,
—	that is, there are eight times as many eighths
32	as whole ones. Hence, in four whole ones,
7	there are eight times four eighths, or thirty-
—	two eighths, and the seven eighths added give
39 $\frac{39}{8}$	thirty-nine eighths.

Hence, to change a mixed number to an equivalent fraction:

Multiply the whole number by the denominator of the fraction; to that product add the numerator, and write the result over the denominator.

To Multiply a Fraction by a Whole Number.

$\frac{3}{7}\times 5$, $\frac{3\times 5}{7}$ I wish to multiply a fraction by a whole $\frac{3}{8}\times 2$, $2)\frac{3}{8}=\frac{3}{4}$. number. The value of a fraction is the quotient arising by dividing the numerator by the denominator. If the numerator be increased by any factor, the number of times that it will contain the denominator will be proportionally increased; hence the value of the fraction will be proportionally increased.

Again—If the denominator be diminished by any factor, the number of times that it will be contained in the numerator will be proportionally *increased*; hence, the value of the fraction will be proportionally increased.

Therefore, to multiply a fraction by a whole number:
Multiply the numerator, or *divide* the *denominator*, by the whole number.

To Multiply a Whole Number by a Fraction.

$5\times\frac{3}{4}$ $\frac{5\times3}{4}$
5×3

I wish to multiply a whole number by a fraction; for instance, 5 by $\frac{3}{4}$, or by one-fourth part of three. First, I will multiply by the whole of three. (5×3.) Now as the multiplier has been four times as large as it should be, the *product* is four times as large as *it* should be; hence to make it what it should be, I will divide it by four, represented thus: $\frac{5\times3}{4}$.

Hence, to multiply a whole number by a fraction:
Multiply the whole number by the numerator of the fraction, and divide that product by the denominator.

To Divide a Fraction by a Whole Number.

$3)\frac{6}{11}$ $3)\frac{6}{11}=\frac{2}{11}$
$4)\frac{5}{9}$ $\frac{5}{9\times4}$

I wish to divide a fraction by a whole number. The value of a fraction is the quotient arising by dividing the numerator by the denominator. If the numerator be diminished by any factor, the number of times that it will contain the denominator will be proportionally diminished; hence the value of the fraction will be proportionally diminished.

Again.—If the denominator be increased by any factor, the number of times that it will be contained in the numerator will be proportionally *diminished;* hence, the value of the fraction will be proportionally diminished.

Therefore, to divide a fraction by a whole number:
Divide the numerator, or *multiply* the *denominator* by the whole number.

To Divide a Whole Number by a Fraction.

$\frac{3}{4})5$, $\frac{5}{3}$,
$\frac{5\times4}{3}$,

I wish to divide a whole number by a fraction; for instance, 5 by $\frac{3}{4}$, or by one-fourth part of three. First, I will divide by the *whole* of three. ($\frac{5}{8}$.) Now as the divisor has been four times as large as it should be, the quotient ($\frac{5}{8}$) is *one-fourth as large* as *it* should be; hence, to make it what it should be, I will multiply it by four, ($\frac{5\cdot4}{8}$) which is done by multiplying the numerator by four, as before explained.

Hence, to divide a whole number by a fraction:

Multiply the whole number by the denominator of the fraction and divide that product by the numerator.

To Multiply one Fraction by another.

$\frac{3}{4}\times\frac{5}{7}$, $\frac{3\cdot5}{4}$, $\frac{3\times5}{4\times7}$,

I wish to multiply one fraction by another; for instance, $\frac{3}{4}$ by $\frac{5}{7}$, or by one-seventh part of five. First, I will multiply by the whole of five, ($\frac{3\cdot5}{4}$,) which is done as before explained, by multiplying the numerator by five. Now as the multiplier has been seven times as large as it should be, the *product* is seven times as large as *it* should be; hence, to make it what it should be, I will divide it by seven, which is done as before explained, by multiplying the *denominator* by seven, ($\frac{3}{4}\cdot\frac{5}{7}$.) I notice that the numerators have been multiplied together for a new numerator, and the denominators have been multiplied together for a new denominator.

Hence, to multiply one fraction by another:

Multiply the numerators together for a new numerator, and the denominators together for a new denominator. If there are mixed numbers, first change them to improper fractions.

To change Compound Fractions to Simple ones.

$\frac{3}{7}$ of $\frac{5}{8}$, $\frac{5}{7\times8}$, $\frac{3\times5}{7\times8}$,

I wish to reduce compound fractions to simple ones, or to find what *of* means between two fractions; for instance, I wish to obtain $\frac{3}{7}$ of $\frac{5}{8}$. First, I will find *one*-seventh of five-eighths, which is done by dividing by seven, and *that* is done by multiplying the *denominator* by seven, as before explained. ($\frac{5}{7\cdot8}$.) Now I have one-seventh of five-eighths. Three-sevenths will evidently be obtained by multiplying one-seventh by three, and *that* is done by multiplying the *numerator* by three, as before explained, ($\frac{3}{4}$; $\frac{5}{8}$.) I notice that I have multiplied the numerators together, and the denominators together—the same as in multiplication of one fraction by another.

Therefore, we may conclude, that *of* occuring between two fractions shows that they are to be multiplied together.

To Divide one Fraction by another.

$\frac{5}{9})\frac{3}{4}$,

$\frac{3}{4\times5}$, $\frac{3\times9}{4\times5}$,

I wish to divide one fraction by another. For instance, $\frac{3}{4}$ by $\frac{5}{9}$ or by one-ninth part of five. First, I will divide by the whole of five, which is done as before explained, by multiplying the denominator by five. ($\frac{3}{4.5}$.) Now as the divisor has been nine times as large as it should be, the quotient is only one ninth as large as *it* should be ; hence, to make it what it should be, I will multiply it by nine, which is done as before explained, by multiplying the numerator by nine. ($\frac{3}{4}$: $\frac{9}{5}$.)I notice that in the operation the terms of the divisor have been inverted, and it is then multiplied into the dividend.

Hence, to divide one fraction by another :

Invert the terms of the divisor, and then proceed as in multiplication of one fraction by another. If there are mixed numbers, first change them to improper fractions.

ADDITION OF VULGAR FRACTIONS.

$\frac{7}{18}$, $\frac{5}{12}$, $\frac{4}{9}$, $\frac{3}{5}$.

I wish to add $\frac{7}{18}$, $\frac{5}{12}$, $\frac{4}{9}$ and $\frac{3}{5}$, together. As the fractions have not the same denominator, it is evident that they cannot be added unless they shall be changed to equivalent fractions having a common denominator.

As the numerator of a fraction is always such a part of the denominator as the value of the fraction is intended to represent, it is evident that the common denominator must be such a number as will contain each of the given denominators. Such a number as will contain each of the denominators, is a common multiple of them. And, that the number used may be as small as possible, we will use for a common denominator, the least common multiple of all the denominators. The least common multiple of 18, 12, 9 and 5, is 180. Hence, 180 is the least common denominator required. To obtain the numerators, take such a part of the common denominator as the value of each fraction represents. For the first numerator, take $\frac{7}{18}$ of 180. One eighteenth of 180 is 10, and seven-eighteenths of 180 is 70. Hence, the first equivalent fraction is $\frac{70}{180}$—the second is $\frac{75}{180}$—the third is $\frac{80}{180}$—and the fourth is $\frac{108}{180}$. Now, as the fractions have a common denominator, they may be added by adding their numerators, and writing the result over the common denominator.

Hence, to add fractions :

Find the least common multiple of all the denominators, for a common denominator—take such parts of that common denominator for a numerator, as the values of the given fractions represent—then add the numerators, and write their sum over the common denominator. Compound fractions must be reduced to simple ones.

SUBTRACTION OF FRACTIONS.

As subtraction is the reverse of addition, we have the following rule for the subtraction of fractions :

Change the given fractions to equivalent fractions having a common denominator, and then subtract their numerators, and write the difference over the common denominator.

DECIMAL FRACTIONS.

$\frac{5}{8}$	8)5000	.6
	625	.62
	1000	.625

I have the expression $\frac{5}{8}$; which shows that 5 is to be divided by 8. As 5 will not contain 8, I will annex a naught to the 5, which is equivalent to multiplying it by ten, and then I will divide.

I obtain 6 for a quotient, and a remainder of 2. Now as the dividend is ten times as large as it should be, the *quotient* (6) is ten times as large as *it* should be; hence, to make it what it should be, I will divide it by ten. To continue the division farther, I will annex another naught, equivalent to multiplying the original dividend by 100. I now obtain 2 for a quotient, and a remainder of 4. Now, as the dividend is one hundred times as large as it should be, the *quotient* (62) is one hundred times as large as *it* should be; hence, to make it what it should be, I will divide it by 100. To continue the division still farther, I will annex another naught; equivalent to multiplying the original dividend by one thousand. I now obtain 5 for a quotient, and I have no remainder. Now as the dividend is one thousand times as large as it should be, the *quotient* (625) is one thousand times as large as *it* should be; hence, to make it what it should be, I will divide it by 1000.

I notice that when I had one figure in the numerator of the quotient, I had for a denominator one naught and a one;—when I had two figures in the numerator, I had *two* naughts and a one for a denominator;—when I had three figures in the numerator, I had three naughts and a one for a denominator. And I have here such a set of fractions as, in general, have for a denominator 1, followed by as many naughts as there are figures in the numerator.

Mathematicians have agreed that to represent such denominators, they will write a point at the left of the numerator. Thus .6 (six with a point at the left,) has a denominator of 10, and hence is read six-tenths. .62 (sixty-two with a point at the left,) has a denominator of 100, and hence is read sixty-two hundredths. .625 (six hundred and twenty-five with a point at the left,) has a denominator of 1000, and hence is read six hundred and twenty-five thousandths. Tenths,—hundredths,—thousandths,—fractional expressions whose values increase (from right to left) in a tenfold proportion, and hence they are called *decimal fractions.*

It is plain, that the farther a figure is removed to the right of the decimal point, the less its value. Hence the placing of a naught between the decimal point and the first significant figure, renders the value of the expression only one tenth as great. Again, as a naught annexed has no value *itself*, and as it does not change the distance of the other figures from the decimal point, it will not affect the value of the expression.

As the first figure at the right of the decimal point is of one-tenth as great value as units, it is evident that decimal fractions may be annexed to whole numbers, as one continuous number, if we observe to place a point between the tenths and units. Such expressions are called mixed numbers.

ADDITION AND SUBTRACTION OF DECIMALS.

To add or subtract decimal fractions—as they increase or decrease as whole numbers, we must write them so that the decimal points, and units of the same denomination, will fall under each other; and then proceed as in whole numbers.

MULTIPLICATION OF DECIMALS.

I wish to multiply one decimal fraction by another, and to find what number of figures to point off at the right hand of the product, for decimals.

For instance I wish to multiply 2.35
by .7

1.645

From the theory of decimals, before explained, I know that the 235 may have a denominator of a one and two naughts, ($\frac{235}{100}$) and that the 7 may have a denominator of a one and one naught. ($\frac{7}{10}$). Here then I have them expressed in the form of *vulgar fractions.* ($\frac{235}{100}\times\frac{7}{10}$). But to multiply one vulgar fraction by another, as before explained, we multiply the numerators together for a new numerator, and the denominators together for a new denominator. ($\frac{1645}{1000}$).

Now there will always be as many naughts at the right hand of the denominator of the product, as there are at the right of the denominators of both multiplier and multiplicand,—but, as the number of naughts shows the number of *decimal* figures, there will always be as many decimal figures in the product, (if the denominator be removed,) as there are in both multiplier and multiplicand—in this case, three. Hence, to multiply one decimal fraction by another :—

Proceed as in whole numbers and point off at the right of the product, as many figures for decimals, as there are in both of the factors.

DIVISION OF DECIMALS.

.7).245 — .35 I wish to divide one decimal fraction by another, and to find what number of places to point off at the right of the quotient for decimals.

For instance I wish to divide .245
by .7

From the theory of decimals, before explained, I know that the 245 may have a denominator of a one and three naughts, ($\frac{245}{1000}$)—and that the 7 may have a denominator of a one and one naught. ($\frac{7}{10}$). Here then I have them expressed in the form of vulgar fractions. ($\frac{245}{1000} \div \frac{7}{10}$). But to divide one vulgar fraction by another, as before explained, invert the divisor, and then proceed as in multiplication. $\frac{245}{1000} \times \frac{10}{7}$, $\frac{35}{100}$; .35. Now the one naught from the denominator of the divisor will cancel one naught from the denominator of the dividend. If there had been two naughts in the denominator of the divisor, they would have cancelled two naughts from the denominator of the dividend. And there will always be as many naughts left at the right for the denominator of the quotient, as those at the right of the denominator of the dividend, exceed those in the denominator of the divisor. But, as the number of naughts shows the number of decimal figures, there will be as many decimal figures in the quotient, (if the denominator be removed,) as those in the denominator of the dividend exceed those in the denominator of the divisor.

Hence, to divide one decimal fraction by another :

Proceed as in whole numbers, and point off, at the right of the quotient, as many figures for decimals, as those in the dividend exceed those in the divisor. If there are less decimal figures in the dividend than in the divisor, annex naughts to the dividend.

CANCELLATION.

It has already been shown that if we multiply and then divide the product by the same number, the value of the original number will not be changed. Hence, it will appear evident that if I wish to multiply 386 by 3 and 4, and then divide it by 3 and 2, that to multiply it by 2 would produce the same result. This operation may be expressed thus :

$\not{3}$	386
	$\not{3}$
$\not{2}$	$\not{4}$
	772 Prod.

Multiply 872 by 16, 25 and 8, and divide by 32, 6, 5 and 20 ; and what number will result ?

Operation

$\not{3}\not{2}$	872
$\not{6}$	$\not{1}\not{6}$
$\not{5}$	$\not{1}\not{2}$
$\not{2}\not{0}$	$\not{2}\not{5}$
$\not{2}$	$\not{8}$
$\not{4}$	$\not{5}$
	2
	1744 Prod.

Multiply 75 by 18, 28, 36 and 50, and divide by 25, 72, 84 and 54.

25	75
72	18
84	28
54	36
4	50
6	3
2	9
	25

25 Result.

Multiply 168 by $\frac{3}{4}$, 18, $\frac{5}{9}$, and divide by $\frac{4}{7}$, $\frac{1}{9}$ and 36.

	168
4	3
	18
9	5
4	7
1	9
36	42
2	21

4) 2205

551¼ Ans.

I wish to multiply 168 by one-fourth part of 3. If I place the factor 3 on the right hand side, I make the dividend too large by a factor 4. Hence, that the quotient may be of proper value, I will increase the divisor by a factor 4.

Again, in relation to the $\frac{4}{7}$, if I divide by the whole of 4 it will give a quotient only one-seventh as large as it should be; hence, to make it what it should be, I must increase it by a factor 7, which is done by writing a factor 7 on the dividend side of the line. By a course of reasoning entirely similar, we shall find that in case of fractions, we must *write the numerator on the same side of the perpendicular line that it would be if it were a whole number, and the denominator on the other side.*

Remark.—The active teacher will call the attention of the student to such abbreviations as they may occur, more profitably than it can be done here. And as many opportunities will be offered in the following pages, for practice in this particular, we deem it unnecessary to increase the number of examples.

Examples to be solved by Analysis.

1. If 4 yards cost $7, what will 20 yards cost? Ans. $35.

Note.—20 yards being 5 times as much as 4 yards, they will cost 5 times as much: hence, to find the cost of 20 yards, multiply the cost of 4 yards by 5.

2. If 28 lbs. butter cost $5 92, what cost 7 lbs.? Ans. $1 48.

Note.—7 lbs. being $\frac{1}{4}$ of 28 lbs., they will cost only $\frac{1}{4}$ as much: hence, to find the cost of 7 lbs., divide the cost of 28 lbs. by 4.

Remark.—Let the teacher compel the student to give a good reason for each step.

Appeal.—If you are a student, do not pass over a point without knowing fully all the reasons why, and then learn to tell those reasons to others fluently.

3. If 4½ tons of hay will keep 3 cattle during the winter, how many tons will it take to keep 25 cattle the same time? Ans. 37½.

Note.—First find the amount required to keep one, and then multiply by 25.

4. What will 9 yards of cambric cost, if $40 96 be paid for 72 yards? Ans. $5 12.

5. If 12½ lbs. of sugar cost $1 34, what will 1 cwt. cost? Ans. $10 72.

6. If a bankrupt pay 12s. 6d. on a pound, how much will A. receive to whom he owes £1000? Ans. £625.

7. What cost 3 hhd. of wine at 10 cts. a pint? Ans. $151 20.

8. If a man performs a journey in 6 days, when the days are 8 hours long, in how many days will he do it when the days are 12 hours long? Ans. 4.

9. If a pipe empty a cistern in 15 hours, how many pipes will empty the same cistern in $\frac{3}{4}$ of an hour? Ans. 20.

10. If 12 pears are worth 21 apples and three apples cost a cent, what will be the price of four-score and four pears? Ans. 49 cents.

11. How long should a person keep $450, to compensate for the use of $150 for 18 months? Ans. 6 months.

12. A man bought 18 pipes of wine at 12s. 6d., New Jersey currency, a gallon; how many dollars did it cost? Ans. $3780.

13. If in four months I spend as much as I gain in three, how much do I lay up at the year's end, if I gain every 6 months $428 50? Ans. $214 25.

14. If 2 lbs. of sugar cost 25 cents, and 8 lbs of sugar are worth 5 lbs. of coffee, what will 100 lbs. of coffee cost? Ans. $20.

15. How many gallons of water must I add to a pipe of wine for which I gave $105, to reduce the first cost to 75 cents a gallon? Ans. 14 gallons.

16. If I buy 54 barrels of flour for $297, what will 18 barrels cost at the same rate? Ans. $99.

17. If 14 men can reap 84 acres in 6 days, how many men must be employed to reap 44 acres in 4 days? Ans. 11 men.

18. If by working 9 hours a day, 5 men can hoe 18 acres of corn in 4 days, how many acres can be hoed at that rate in 3 days, working 10 hours a day? Ans. 27 acres.

19. If 8 men can build a wall 15 rods long in 10 days, how many men will it take to build a wall 45 rods long in 5 days? Ans. 48 men.

20. A man having $100, spent a part of it, and afterward received five times as much as he had spent, and then his money was double what it was at first. How much did he spend? Ans. $25.

21. A merchant bought a number of bales of velvet, each containing $129\frac{17}{27}$ yards, at the rate of $7 for 5 yards, and sold them out again at the rate of $11 for 7 yards,

and gained $200 by the bargain; how many bales were there? Ans. 9 bales.

PROPORTION.

Suppose that two yards of cloth cost six dollars, and I wish to find what nine yards of the same cloth will cost.

If two yards of cloth cost six dollars, one yard, which is one half of two yards, will cost one half of six dollars, or three dollars. If one yard cost three dollars, nine yards, which is nine times one yard, will cost nine times three dollars, or twenty-seven dollars. Therefore, if two yards of cloth cost six dollars, nine yards of the same cloth will cost twenty-seven dollars.

Now, it is evident that I may always find the cost of any quantity, as I have done in this case, by first obtaining the cost of a unit, and then multiplying that by the number of units. But as dividing to find the cost of a unit will often give an awkward fraction, I will seek some method of solving such problems which shall obviate the necessity for such division.

The first number of yards multiplied by the price per yard will give its cost, and the second number of yards multiplied by its price per yard will give its cost. But the price per yard being the same in both cases—calling these terms—the first term is multiplied by the same number to make the second, that the third term is multiplied by to make the fourth,—that is, the same ratio* exists between the first and second terms, that there is between the third and fourth. And whenever we have four terms which have the same ratio between the first and second that there is between the third and fourth, they are said to be in proportion. And to show that such relation exists between the terms, dots are used as you see here.

*Ratio is the quotient arising from dividing one number by another.

yds.		$		yds.		$
2	:	6	: :	9	:	27

The first and last terms are called *Extremes*, and the second and third terms are called *Means*. The first and second taken together are called a couplet, and the third and fourth are called a couplet. The first term in each couplet is called the *antecedent*, and the second term in each couplet is called the *consequent*.

I will now use the terms of the first couplet as multiplicands. Now as the second multiplicand is three times the first, if I obtain equal products, it is evident that the first multiplicand must be multiplied by a number three times as large as that by which we must multiply the second. If the second multiplicand had been five times the first, then the first must have been multiplied by a number five times as large as that by which we multiply the second. If the second had been nine times the first, then the first must have been multiplied by a number nine times as large as that by which we multiply the second. And whatever ratio exists between the multiplicands must exist between the multipliers, in an inverse order. But from the nature of Proportion, the same ratio always does exist in the last couplet, that there does in the first; hence, if the second couplet be inverted and multiplied into the first, we shall always obtain two equal products; that is, the product of the means *is* always equal to the product of the extremes.

This being the case, if a factor in one of the extremes is wanting, as in the case above, we may multiply together the factors of the means, and divide their product by the product of the given factors of the extremes. Or if a factor in the means is wanting, multiply together the factors of the extremes, and divide their product by the product of the given factors of the means.

On the principle that like causes produce like effects, and unlike causes produce corresponding unlike effects, it is evident that if in arranging the terms, the first and third be made to represent the cause of the effects which the second and fourth terms represent, there will be the same ratio between the first and second terms as between the third and fourth. Hence, the rule:

State the question, by writing that which represents the cause of one effect in the first term, and its effect in the second term; and that which represents the cause of the other effect in the third term, and that which represents the effect itself in the fourth term; writing some character to represent the unknown quantity. Then cause the corresponding factors in the first and third terms to be of the same denomination, also the corresponding factors of the second and fourth terms. Then if the unknown factor fall in either extreme, divide the product of the means by the product of the given factors of the extremes. And if the unknown factor fall in either of the means, divide the product of the extremes by the product of the given factors of the means.

EXAMPLES IN PROPORTION.

1. If 6 horses consume 34 bushels of oats in a certain time, how many bushels will 18 horses consume in the same time? Ans. 102 bu.

C.		E.		C.		E.
6	:	34	: :	18	:	x

The statement reads—If 6 horses consume 34 bushels in a certain time, 18 horses will consume how many bushels? As the unknown factors falls in the extremes, the extremes will form the divisor and the means the dividend, thus:

C.		E.		C.		E.
6	:	34	: :	18	:	x

Divisor	Dividend
~~6~~	34
	~~18~~
	3

Apply the principles of cancelation, as before explained, and we have on the dividend side 34 to be multiplied by 3, which gives 102—hence, if 6 horses consume 34 bushels of oats in a certain time, 18 horses will consume 102 bushels in the same time.

2. If a tree 20 feet high cast a shadow 30 feet long, how long will be the shadow of a tree 50 feet high?

Ans. 75 feet.

C.		E.		C.		E.
20	:	30	: :	50	:	x

The statement reads—If 20 feet of tree cause 30 feet of shadow, 50 feet of tree will cause how much of shadow? The extremes are the divisor and the means the dividend, &c. The statement might, perhaps, with equal propriety, have read—if 30 feet of shadow require 20 feet of tree, how many feet of shadow will require 50 feet of tree? The means would then have been the divisor and the extremes the dividend—but of course the same result would have been obtained.

C.		E.		C.		E.
30	:	20	: :	x	:	50

3. What will 20 yards of cloth cost, if 6 yards can be bought for $30? Ans. $100.

C.		E.		C.		E.
6	:	30	: :	20	:	x

If 6 yards cost $30, 20 yards will cost how much?

4. Bought 18 barrels of flour for $72; what will be the cost of 42 barrels at the same rate? Ans. $168.

C.		E.		C.		E.
18	:	72	: :	42	:	x

If 18 barrels cost $72, 42 barrels will cost how much?

5. If 8 hats cost $24, what will 110 cost at the same rate? Ans. $330.

6. If 3 barrels of flour cost $24, what will 39 barrels cost? Ans. $312.

7. If 16 pounds of sugar cost $2 56, how much will 13 pounds cost? Ans. $2 08.

8. If a man travel 105 miles in 6 days, how far will he travel in 120 days? Ans. 2100 miles.

9. If 6 yards of cloth cost 18 shillings, what will 11 yards cost? Ans. 33 shillings.

10. If 5 barrels of salt cost $7 25, what will be the cost of 18 barrels? Ans. $26 10.

11. What will 16 tons of hay cost if 6 tons cost $21? Ans. $56.

12. If 6 bushels of corn cost $4 75, what will 75 bushels cost? Ans. $59 37.5.

13. If 7 yards of cloth cost $15 47 what will 12 yards cost? Ans. $26 52.

14. What cost 36 pounds of cloves at 8 cents for 5 oz.? Ans. $9 21$\frac{3}{5}$.

REMARK.—In this example the 36 pounds must be reduced to ounces, to agree with the other denomination of weight.

15. If 2 cords of wood cost $11 50, what will 17 cords cost? Ans. $97 75.

16. If $4 75 be paid for 19 lbs. of salmon, how many pounds will $25 50 buy? Ans. 102.

17. If 7 ounces of silver are sufficient to make 17 teaspoons, how many may be made from 21 pounds? Ans. 816.

18. If 5 and 2 make 10, how much will 9 and 5 be? Ans. 20.

19. If 4$\frac{1}{2}$ yards cost $9 75, what will 13$\frac{1}{2}$ yards cost? Ans. $29 25.

C.	E.	C.	E.
$4\frac{1}{2}$:	9 75 ::	$13\frac{1}{2}$:	x.
$\frac{9}{2}$:	9 75 ::	$\frac{27}{2}$:	x

Notice, the 9 must be written on the divisor side, and the 2 (its denominator,) on the dividend side of the perpendicular. All mixed numbers must first be reduced to improper fractions. For suppose that, in this example, we write the 4 on the divisor side, and the 2 on the other side; our divisor is then only $\frac{1}{2}$ of 4, but it should be $\frac{1}{2}$ of 9.

20. If 6$\frac{1}{2}$ bushels of oats cost $3, what will 9$\frac{1}{4}$ bushels cost? Ans. $4 26.9.

$$
\begin{array}{ccccccc}
C. & & E. & & C. & & E. \\
6\frac{1}{2} & : & 3 & :: & 9\frac{1}{4} & : & x \\
\frac{13}{2} & : & 3 & :: & \frac{37}{4} & : & x
\end{array}
$$

$$
\begin{array}{r|l}
13 & \not{2} \\
 & 3 \\
\not{4} & 37 \\
2 &
\end{array}
$$

21. If $\frac{3}{5}$ of an ell-English cost $\frac{1}{3}$ of a pound, what will $12\frac{1}{2}$ yards cost? Ans. £5 11*s.* $1\frac{1}{3}$*d.*

$$
\begin{array}{ccccccc}
C. & & E. & & C. & & E. \\
\frac{3}{5} & : & \frac{1}{3} & :: & \frac{25}{2} & : & x \\
5 & & & & 4 & &
\end{array}
$$

NOTE. If $\frac{3}{5}$ of an ell-English of 5 quarters each cost £$\frac{1}{3}$, $\frac{25}{2}$ yards of 4 quarters each cost how much?

22. If $12\frac{1}{2}$ cwt. cost \$$42\frac{1}{4}$, what will $48\frac{3}{8}$ cwt. cost? Ans. \$163 50.

It will appear evident that we may reduce the vulgar fractions to equivalent decimals.

$$
\begin{array}{ccccccc}
C. & & E. & & C. & & E. \\
12.5 & : & 42.25 & :: & 48.375 & : & x
\end{array}
$$

And now if we will make the decimal places in the two causes equal, the decimal points may be omitted. For suppose we allow them to remain, and we consider the fractions to have denominators expressed—

$$
\frac{125}{10} : \frac{4225}{100} :: \frac{48375}{1000} : x
$$

$$
\begin{array}{r|l}
125 & \not{1}\not{0}\not{0} \\
\not{1}\not{0}\not{0} & 4225 \\
1000 & 48375
\end{array}
$$

It will be seen that when the decimal figures are thus made equal, they exactly balance against each other, (as the two one hundreds in this case,) and hence the decimal points may be omitted.

23. If $\frac{1}{2}$ of a yard cost $\frac{3}{5}$ of a pound, what will $\frac{7}{8}$ of a yard cost? Ans. £1 1*s.*

24. If $\frac{7}{16}$ of a ship cost £51, what are $\frac{3}{32}$ of her worth? Ans. £10 18*s.* $6\frac{6}{7}$*d.*

25. If $9\frac{5}{7}$ yards cost $\$11\frac{3}{5}$, what will $16\frac{16}{203}$ ells-English cost? Ans. $24.

26. If $4\frac{1}{2}$ tons of hay feed 3 cattle during the winter, how many tons will keep 25 cattle the same time?

Ans. $37\frac{1}{2}$.

27. What cost 1 cwt. of cheese at 24 cents for $3\frac{1}{2}$ lbs.?

Ans. $\$6\ 85\frac{5}{7}$.

28. If $\frac{3}{8}$ of a ship is worth $6000, how much is $\frac{5}{16}$ of her worth? Ans. $5000.

29. If $5\frac{1}{2}$ yards of cloth are worth $\$27\frac{1}{2}$, how much are $50\frac{1}{4}$ yards worth? Ans. $251 25.

30. How much cotton cloth can I buy for £10 6*s.* 8*d.*, if I pay $6\frac{1}{4}$ pence for $1\frac{1}{4}$ yards? Ans. 496 yards.

31. If 4 men can mow 12 acres of grass in 3 days, how many acres can 8 men mow in 18 days? Ans. 144.

$$\begin{matrix} m4 & & & m8 & & \\ d3 & : & 12 & :: & d18 & : & x \end{matrix}$$

If 4 men in 3 days mow 12 acres of grass, 8 men in 18 days will mow how many acres?

Direction.—Arrange both antecedents in the same order, and both consequents in the same order.

Remark.—It will be observed that animal strength and time are most conveniently considered as the cause; money and time also are often most conveniently considered as cause.

32. If 4 students spend $96 in 3 months, how much will 6 students spend in 10 months, at the same rate?

Ans. $480.

33. If 20 horses eat 18 bushels in 10 days, how many bushels will feed 50 horses 20 days? Ans. 90.

34. If 18 men in 40 days, working 9 hours a day, can build a wall 240 feet long, 8 feet wide, and 6 feet high; in how many days, of 12 hours in a day, can 6 men build a wall 160 feet long, 6 feet high and 4 feet wide?

Ans. 30.

35. If 6 lbs. of cotton make 20 yards of cloth, an ell French wide, how many pounds will be required to make a piece of cloth 150 yards long, and an ell-Flemish wide?

Ans. $22\frac{1}{2}$.

36. How much money will gain $6 in 3 months, if $350 in 9 months gain $15? Ans. $420.

37. If 45 yards of cloth, 6 quarters wide, will make 10 suits of clothes, how many pieces, each 45 yards in length and 3 quarters wide, will make 500 suits of clothes?
Ans. 100.

38. If a cellar $22\frac{1}{2}$ feet long, $17\frac{3}{10}$ feet wide, and $10\frac{1}{4}$ feet deep, be dug in $2\frac{1}{2}$ days by 6 men working $12\frac{3}{10}$ hours in a day, how many days of $8\frac{1}{5}$ hours each, should 9 men work, to dig another measuring 90 feet long, $34\frac{3}{5}$ feet wide, and $12\frac{3}{10}$ feet deep? Ans. 24 days.

39. A trench of 5 degrees of hardness, 70 yards long, 3 yards wide, and 2 yards deep, can be dug by 336 men, in 5 days of 10 hours each; now, 240 men, in 9 days, can dig how long a trench of 6 degrees of hardness, 5 yards wide, and 3 deep, if they work 12 hours each day?
Ans. 36 yards.

40. If 3 men receive $18 for working 3 days, of 10 hours each, how much should 8 men receive for working 9 days of 12 hours each? Ans. $172 80.

41. If 22 men can cut 81 cords of wood in 3 days, when the days are 14 hours long, how many men will be required to cut 162 cords in 4 days, when the days are 11 hours long? Ans. 42 men.

42. If a stream of water running into a pond of 190 acres, will raise the pond 10 inches in 12 hours, how much would a pond of 50 acres be raised by the same stream, in 10 hours? Ans. $31\frac{2}{3}$.

43. If 175 bushels of corn, when corn is 60 cents a bushel, be given for the carriage of 200 barrels of flour 58 miles,—when corn is worth 75 cents a bushel, how much must be given for the carriage of 90 barrels of flour, 200 miles? Ans. $434\frac{14}{29}$ bu.

44. A garrison of 900 men have provisions for 4 months; how many must leave, that the provisions may last the remainder 9 months? Ans. 500.

45. If a cistern $7\frac{2}{3}$ feet long, $5\frac{2}{5}$ feet wide, and $4\frac{1}{2}$ feet deep, hold 784 measures of water, how many measures

would it hold if the length were doubled, the width made 3 times as great, and the depth 4 times as great?

Ans. 18816.

46. If $\frac{2}{3}$ of a job may be performed by 12 men laboring 7 hours a day for $15\frac{5}{9}$ days, how many men may be withdrawn and the residue of the job be finished in 15 days more, if the laborers are employed 7 hours a day?

Ans. 4 men.

47. In what time will $627 50, loaned at 7 per cent, produce as much interest as $2510, at $3\frac{1}{2}$ per cent, will produce in 1 year and 8 months? Ans. $3\frac{1}{3}$ years.

48. If 1000 men, beseiged in a town with provisions for 5 weeks, allowing each man one pound a day, be reinforced by 500 men more; and supposing that they cannot be relieved until the end of 8 weeks; how many ounces a day must each man have, that the provisions may last that time? Ans. $6\frac{2}{3}$ ounces.

49. If 6 is 10, how much is 12? Ans. 20.

50. If 12 oxen graze down 16.25 acres of grass in 20 days, how many acres of the same pasture will be sufficient for 24 oxen 24 days? Ans. 39.

51. If 3 masters, who have each 8 apprentices, in 5 weeks of 6 days each, earn $360, how much will 5 masters, who have each 10 apprentices, earn in 8 weeks, each week $5\frac{1}{2}$ days, their daily wages being the same, the masters working as well as the apprentices?

Ans. $1075 $55\frac{5}{9}$.

52. How many laborers must be employed to finish a piece of work in 15 days, which 5 can do in 24 days?

Ans. 8.

53. If when wheat is 83 cents a bushel, the cent loaf weighs 9 ounces, what should it weigh when wheat is $1 24.5 a bushel? Ans. 6 oz.

54. If 9 yards of cloth cost £5 12*s*., how many yards of cloth can be bought for £44 16*s*.? Ans. 72.

55. What will 26 yards of cloth come to, if $6 90 be paid for 13 ells-French? Ans. $9 20.

56. A certain piece of corn may be hoed in 9 days by a man who can hoe 5 rows in the same time that each of two

boys can hoe 3; in how many days will the field be hoed, if they all work together? Ans. $4\frac{1}{11}$ days.

57. A can mow an acre of grass in 6 hours, and B in 8 hours; how much will they jointly mow in 10 hours?

Ans. $2\frac{11}{12}$.

58. If A and B can do a piece of work in 7 days, and B alone in 12 days, in how many days can A alone do $\frac{2}{3}$ of it? Ans. $11\frac{1}{5}$.

59. If 7 oz. 5 pwt. of bread be bought for $4\frac{3}{4}d.$ when corn is 4*s.* 2*d.* per bushel, what weight of it may be bought for 1*s.* 2*d.*, when the price per bushel is 5*s.* 6*d.*?

Ans. 1lb. 4oz. $3\frac{477}{627}$pwts.

60. If the freight of 9 hhds. of sugar, each weighing 12cwt., 20 leagues, cost £16, what must be paid for the freight of 50 tierces, each weighing $2\frac{1}{2}$cwt., 300 miles?

Ans. £92 11*s.* $10\frac{2}{9}d.$

INTEREST.

Interest is an allowance made, at a given rate by the hundred, for the use of any commodity whatever.

For instance, if an individual hires one hundred bushels of grain, and pays six bushels for the use of it, he pays an interest of six bushels on one hundred, or six hundredths part of the quantity hired. And whatever is the quantity hired, if he pays a quantity equal to six one-hundredths of it for the use of it, he is said to pay an interest of six per cent.

If he pays this for the use of that which he hired for one year, he is then said to pay an interest of six per cent per annum. If he pays an interest equal to seven one-hundredths part of that which he hires, he is then said to pay an interest of seven per cent per annum, &c.

The quantity hired is called the *Principal.*

At 6 per cent per annum, the interest for one year is = .06 of the Principal.
. . . . month is =.005 . . . .
. . . . 6 days is =.001 . . . .

I wish to find the interest of $675 48, for 2 yrs. 3 mo. 18 days, at 6 per cent per annum.

If for one year the interest is .06 part of the principal, for two years it will be twice as much or .12 part of the principal. If for one month the interest is .005 part of the principal, for 3 months it will be three times .005 or .015 part of the principal. If for six days the interest is equal to .001 part of the principal, for eighteen days, which is three times six days, it will be three times .001 or .003 part of the principal. These three results added give .138, as the fractional part of the principal which will be equal to the interest for 2 years 3 months 18 days. Hence, multiply the principal by that and we have the interest.

```
   675.48
     .138
  -------
   540384
  202644
  67548
  -------
 93.21624
```

Suppose the interest had been required at seven per cent per annum. It is evident that if we divide the interest at six per cent by six, we shall have the interest at one per cent. Then if I multiply the interest at one per cent by seven, I shall have the interest at seven per cent.

Hence, to find the interest on any given sum, for any given time :

Multiply .06 by the number of years—then multiply .005 by the number of months—then multiply .001 by the number of times six is contained in the number of days ; add the three products, and then multiply the principal by their sum, and the product will be the interest at six per cent per annum.

To find the interest at any other rate than six per cent : Divide the interest at six per cent by six, and multiply the quotient by the given rate per cent.

Amount is the sum of principal and interest.

Interest upon Notes, on which partial payments have been made.

The following rule evidently brings to a correct result in computing interest on notes, when partial payments have been made.

Find the *amount* of the principal from the time it became due until the time of settlement; and then find the amount of each payment, from the time it was made until settlement, and subtract their sum from the amount of the principal.

1. ROCHESTER, Jan. 1st, 1848.

For value received I promise to pay to John Short, or order, on demand, $1000 with interest from date.

CHARLES PAYIT.

On this note were the following endorsements :

April 1st, 1848, received $250. Aug. 1st, 1848, received $225. March 16th, 1849, received $116. What was due at the time of settlement, June 25th, 1850—interest being reckoned at 6 per cent per annum?

OPERATION.

Principal - - -	$1000	
Interest for 2 years, 5 mo. and 24 da. -	149	
		$1149
First payment - -	$250	
Interest for 2 yrs. 2 mo. 24 da. -	33 50	
Second payment - -	225	
Interest for 1 yr. 10 mo. 24 da. -	25 65	
Third payment - -	116	
Interest for 1 yr. 3 mo. 9 da. -	8 87	
Sum of amounts - - - -	$659 02	

Balance remaining due, June 25th, 1850, $489 98

2. BUFFALO, July 7th, 1842.

On or before the first day of Jan., 1846, I promise to pay James Takeall, or order, $500, with interest at 6 per cent, for value received.

WM. NEVERPAY.

On this note were the following endorsements :

Jan. 7th, 1843, received $25. April 19th, 1843, received $5. Dec. 13th, 1843, received $77. Dec. 25th, 1845, received $200. What was due at the time of settlement? Ans. $297 54.4.

The substance of the following rule for computing the interest on notes, is established by law in many of the States, and by the Supreme Court of the U. S.

Rule.—Compute the interest on the given principal, from the time the note was given to the time of the first payment. If the payment equals or exceeds the interest, add the interest thus found to the principal, and from that amount subtract the payment and use the remainder as a new principal. But if the payment is less than the interest, compute the interest to the date of the second payment, then from the amount, subtract the sum of the payments then made, unless they are together less than the interest now due ; in which case, compute the interest to the time of the third payment, &c. Whenever the payment (or sum of payments) exceeds the interest due, add the interest to the principal, and from the amount subtract the payment (or payments,) and use the remainder for a new principal, with which proceed as before.

1. A bond or note, dated Feb. 1st, 1810, was given for $500, interest at 6 per cent, and there were payments endorsed upon it as follows :

May 1st, 1810 - -	$40
November 14th, 1810 - -	8
April 1st, 1811 - - - -	12
May 1st, 1811 - - -	30

How much remains due on said note, September 16th, 1811? Ans. $455 57.

BOSTON, Jan. 1st, 1841.

2. For value received, I promise to pay Monroe Bogardus, or order, $850, on demand, with interest at 6 per cent.

JONATHAN LOSEITALL.

On this note the following payments were endorsed :

July 1st 1841, received - -	$100 62
Dec. 1st, 1841, received - -	15 28

Aug. 13th, 1842, received - - 175 75

What was due January 1st, 1843 ? Ans. $650 38.

3. For value received I promise to pay Wm. Cunningham, or order, eight hundred sixty-seven dollars and thirty-three cents, with interest at 6 per cent.

JOHN BANKRUPT.

On this note were the following endorsements :

April 16th, 1823,	- -	$136 44
April 16th, 1825,	- -	319
Jan. 1st, 1826,	- - -	518 68

What remained due July 11th, 1827 ?

Ans. $215 10.3.

COMPOUND INTEREST.

When the amount is found at the end of each period of time, and is used as principal for the next period, the interest is said to be compound. Hence for computing Compound Interest, we have the following rule :

Find the amount for the first period of time, (one year, or one half year, as the case may be,) and use it as a new principal, upon which find the amount for the second period of time, and thus proceed through all the different periods required. From the last amount subtract the first principal, and the remainder will be the Compound Interest.

EXAMPLES.

1. What is the compound interest of $500 for 3 years, at 7 per cent ? Ans. $112 52.1.

2. What is the compound interest of $316 for 3 years, 4 months and 18, days at 6 per cent ? Ans. $69 01.7.

3. What will be the compound interest on $500 for one year, at 8 per cent, the interest being computed quarterly? Ans. $41 21.6.

4. What is the amount of $1200, at compound interest for 4 years, at $4\frac{1}{4}$ per cent, interest being due annually ? Ans. $1417 37.7.

5. What is the compound interest of £760 10*s*. for 4 years, at 4 per cent per annum ? Ans. £128 12*s*. 3¼d.

Given, the amount, principal, and time, to find the rate.

The amount is made up of the principal and interest added; hence if I subtract the principal from the amount, I shall have left the interest, which is composed of three factors, principal, rate and time, two of which (principal and time,) I have given, to find the other factor. If I divide the interest, which is the product of principal, rate and time, by the product of principal and time, I shall have the rate per cent. Hence the rule :

Subtract the principal from the amount, and divide the remainder by the product of principal and time.

Given, the amount, principal and rate, to find the time.

If I subtract the principal from the amount there will be left the interest, which is composed of the factors principal, rate and time; two of which, (principal and rate,) are given to find the time. Hence, we have to divide the interest, which is the product of principal, rate and time, by the product of principal and rate, and we have the time. Hence, the rule :

Subtract the principal from the amount, and divide the remainder by the product of principal and rate.

COMMISSION AND INSURANCE.

Commission is an allowance of a certain per cent. for transacting business, such as buying and selling goods, &c.

Insurance is security against loss, by fire or otherwise, and is obtained by paying a certain per centage on what is called the *Premium.*

Premium is a certain per centage on the value of the property insured.

To estimate Commission or Premium :

Multiply the given sum by the given per cent, considered as hundredths.

STOCKS.

Stock is a general term applied to government funds, and to the capital of incorporated companies.

When Stocks sell at their original value, they are said to be at par.

When they sell for more than their original value they are said to be above par.

When they sell for less than their original value they are said to be below par, or at a discount.

LOSS AND GAIN PER CENT.

$$\begin{array}{r} \$5.10 \\ 4.25 \\ \hline .85 \end{array} \qquad \tfrac{85}{425}=.20$$

Suppose a person to buy cloth at \$4 25 per yard, and to sell it \$5 10 per yard, I wish to find what fractional part of the first cost is gained.

If I subtract the buying price from the selling price, the remainder will be the gain on the price of one yard, that is, on \$4 25.

Subtracting, I have a remainder of .85. Now to find what part the .85 is of \$4 25. As the numerator of a vulgar fraction is always such a part of the denominator as the value of the fraction is intended to represent, I have only to write the 85 for a numerator and the 425 for a denominator, ($\frac{85}{425}$.) Now if I change this vulgar fraction to an equivalent decimal, I shall have the same value represented in hundredths—or the gain per cent.

Hence, to find the profit or loss per cent. :

Make the profit or loss the numerator and the buying price the denominator of a vulgar fraction, (having both in the same denomination,) which change to an equivalent decimal.

To find at what price goods must be sold to gain or lose a certain per cent.

$\frac{20}{100}=\frac{1}{5}$, $\frac{5}{5}+\frac{1}{5}=\frac{6}{5}$

```
5 )1.25
   ----
    .25
      6
   ----
  $1.50
```

If I buy wheat at $1 25 per bushel, at what price per bushel must I sell it to gain 20 per cent?

A gain of 20 per cent is equal to $\frac{20}{100}$ of the cost price; which fraction reduced gives $\frac{1}{5}$, hence, I wish to gain $\frac{1}{5}$, therefore I must sell the wheat for $\frac{6}{5}$ of the cost price. $\frac{6}{5}$ of $1 25 is $1 50. Hence, the selling price must be $1 50 a bushel.

Therefore, to find at what price a commodity must be sold to gain or lose at a given per cent:

Make a vulgar fraction of the required per cent, which reduce to its lowest terms. Then add it to or subtract it from a unit, as the case is one of gain or loss, and multiply the given price by the resulting fraction.

NOTE.—If the selling price be given, to find at what price the commodity must have been bought to gain or lose a certain per cent:

Prepare the fraction, as in the rule, and *divide* the given price by it.

DISCOUNT.

Discount is an allowance made for the payment of a sum of money before it becomes due.

```
   1
 .06        1.24 )620.00(500
 ---               620
 .06               ---
   4                 00
 ---
 .24               620
1.00               500
----               ---
1.24               120   Discount.
```

Suppose a person to owe $620, to be paid at the expiration of four years, without interest, and I wish to find what sum may be paid now, without gain or loss—discounting at 6 per cent per annum.

The interest on one dollar for one year is six cents; for four years it is twenty-four cents, and the one dollar being added, gives the amount for four years—one dollar and twenty-four cents. Therefore, one dollar *now*, will be worth one dollar and twenty-four cents, four years from now; and for one dollar and twenty-four cents due four years hence, he can afford to take one dollar now. Hence, as many times as one dollar and twenty-four cents is contained in the $620, so many dollars can be taken now.

```
1 24)620 00(500
     620
     ———
      00
```

Then $500 is the present worth of the $620. And this subtracted from the $620 leaves $120, the amount discounted. Hence, to find the present worth of a sum of money due some time hence without interest:

Find the amount of one dollar for the given rate and time, by which divide the given sum.

NOTE.—To find the discount—subtract the present worth from the original sum.

REMARK.—When Banks discount they reckon the interest upon the given sum in advance, and subtract that, which is evidently more than they should take, by the interest on the true discount.

EQUATION OF TIME.

Suppose an individual to owe $800, to be paid at the expiration of four months, and $800 more, due at the expiration of eight months. I wish to find at what one time the whole may be paid without gain or loss to either party. The present worth of the first $800, is $784 31; and the present worth of the second $800, is $769 23. The sum of these is $1553 54.

Now the question is, in what time will $1553 54 amount to $1600, at 6 per cent per annum?

Subtracting the $1553 54 from $1600, we have $46 64, the interest to gain on $1553 54. If I divide this interest, which is made up of the factors of principal, rate and time, by the product of the two given factors, principal and rate, I shall have the time,

```
                1553 54
                     06
                -------
                93 2124 ) 46 46000 ( .4984 of a year.
                          37 28496
                          --------
                           9175040
                           8389116
 .4984                     -------
    12                      7859240
 -----                      7456992
5.9408                      -------
    30                      4022480
 -----
28.2240
```

Hence the whole should be paid at the expiration of 5 months 28.2 days. Hence the rule:

To find the equaled time of several payments due some time hence without interest:

Find the present worth of each payment separately—then find in what time the sum of these present worths will amount to the sum of the payments.

EXAMPLES.

1. A owes B $107 due at the expiration of 1 year;
   110 50 due at the expiration of $1\frac{1}{2}$ years;
   114 due at the expiration of 2 years.

At what time may the whole be paid without gain or loss to either; reckoning at 7 per cent? Ans. $1\frac{1}{2}$ years.

2. What will be the equated time for the payment of the following bills ?

$103 due in ½ year ;
106 due in 1 year ;
112 due in 2 years ; reckoning at 6 per cent.

Ans. 1 year 1 month 2.99 days.

3. Find the equated time for the payment of the following :

$600 due in 4 years ;
400 due in 5 years ;
220 due in 2 years ;
690 due in 3 years ; reckoning at 5 per cent.

Ans. 3 years 6 months 28.8 days.

4. What would be the difference in time, in the third example if we had reckoned at 6 per cent ?

Ans. 2 months 10 days.

The following rule is in common use, to find the equated time of several payments due at different times :

Multiply each payment by its time, and divide the sum of the products thus obtained by the sum of the payments.

REMARKS.—That this rule does not give correct results is evident from comparing the answers to the foregoing questions, as obtained by the two different rules. The error arises in balancing the interest of a given sum, against the discount of the *same* sum for the same time and rate.

For instance, the equated time for the payment of $800 due in 8 months, and $800 due in 4 months, as found by this rule, is 6 months. That is, the rule seems to suppose the interest on $800 for 2 months equal to the discount of $800 for the same time. Now, whatever is the rate per cent, the *time* as found by this rule, would be the *same.* But is evident that for *higher* rates per cent there will be a *greater* difference between the interest and discount. Hence, the greater the rate per cent, the greater the error by this rule.

1. What would be the answer to the first example, found by the last rule ? Ans. 1 year 6 months 3.6 days.

2. What would be the answer to the second example? Ans. 1 year 2 months 7.6 days.

3. What would be the answer to the third example? Ans. 3 years 7 months 12 days?

4. What difference of time would it make in the fourth example? Ans. 2 months 23.2 days.

SINGLE FELLOWSHIP.

In Fellowship the money invested is called the *Capital* or *Stock*.

The gain is often called the *Dividend*.

Three individuals commence a co-partnership. A puts in \$300; B puts in \$400; C puts in \$500, and they gain \$240. What is each one's share of the gain?

I find the whole amount invested is \$1200; of which A put in $\frac{300}{1200}$, equal to $\frac{1}{4}$ part; B put in $\frac{400}{1200}$, equal to $\frac{1}{3}$ part; C put in $\frac{500}{1200}$, equal to $\frac{5}{12}$ part. Hence A should have $\frac{1}{4}$; B $\frac{1}{3}$; C $\frac{5}{12}$ of the \$240.

Hence, in Fellowship, to find each one's share of the gain or loss:

Make vulgar fractions, having each partner's share of the capital for the numerators, and the whole capital for the denominators, and take such parts of the whole gain or loss as those fractions represent.

DOUBLE FELLOWSHIP.

When time is taken into consideration, we have what is termed Compound Fellowship. For this we have the following rule.

Multiply each man's stock by the time it continues in trade, and use the product as his stock, and form fractions as before.

INVOLUTION.

The Root of a number is one of the equal factors into which the number may be resolved.

The Power of a number is the product arising by using the number some number of times as a factor. Thus:

The 2nd power of 2 is=(2^2)=2×2=4.
" 3rd " " 2 is=(2^3)=2×2×2=8.
" 4th " " 2 is=(2^4)= 16.
" 5th " " 2 is=(2^5)= 32.
" 6th " " 2 is=(2^6)= 64.

The small figures at the right and a little above the root, are called *Indices*, or *Exponents*. They show how many times the root is taken as a factor,—that is, to what power the root is to be raised.

If we multiply 2^3 (the third power of two,) by 2^4 (the fourth power of two,) we shall be combining a multiplier containing four factors of two, with a multiplicand containing three factors of two; hence, the product will contain seven factors of two, or 2^7 (2 seventh power.) And in general; to multiply together different powers of *the same number*, give the number a new exponent equal to the sum of the exponents in both the factors.

Multiply 3^2 by 3^4. 3^6.
Multiply 2^4 by 2^5. 2^9.
Multiply 5^2 by 5^7. 5^9.
Multiply 7^3 by 7^8. 7^{11}.

EVOLUTION, OR EXTRACTING ROOTS.

Table of the second and third powers of the nine digits.

Roots.	1	2	3	4	5	6	7	8	9
Squares.	1	4	9	16	25	36	49	64	81
Cubes.	1	8	27	64	125	216	343	512	729

SQUARE ROOT.

The Square Root of a number is such a number as being multiplied by itself will produce the given number. Thus 3 is the square root of 9, or 6 is the square root of 36.

I wish to find a means of extracting the square root of a number, as 1296. I have first squared each of the nine digits, and I find that none of the squares occupies more than two figures. But as there are more than two figures in this number, (1296,) I know there will be more than one figure in the root. And I now wish to find in what places of the square, the square of each figure in the root will be found.

1	2	3	4	5	6	7	8	9
1	4	9	16	25	36	49	64	81

221	976
200^2=4 00 00	900^2=81 00 00
20^2= 4 00	70^2= 49 00
1^2= 1	6^2= 36

3 6	12.96 (36 Root.
3 6	9
3×3+3×6	66) 3 96
+3×6+6×6	3 96
3^2+2×3×6+6×6	0 00

For that purpose I have taken the two numbers, 221 and 976, and have decomposed them into the parts 200, 20 and 1; and 900, 70 and 6, and have separated each of those parts. From our system of notation I know that the square of the hundreds' figure, (2,) will occupy the fifth place, (4,) counting from the right; and, if the figure is of sufficient value, its square will occupy both fifth and sixth places, (81.) But it cannot occupy more than those two places, as before shown. That the square of the tens' figure, (2,) will occupy the third place from the right, (4,) and, if the figure is of sufficient value, (7,) its square will occupy the third and fourth places, (49.) That the square of the units' figure, (1,) will occupy the first place at the right, and, if the figure is of sufficient value, (6,) its square will occupy the first and second places at the right, (36)—but it cannot occupy more than those two places, as before shown.

Now, as the square of tens cannot reach nearer to the right than the third place, the square of units will be found in the first two places at the right. And as the square of hundreds cannot reach nearer to the right than the fifth place, the square of tens will be found in the next two places; the square of hundreds in the next two places, &c.

Hence, if in the extraction of square root we point the number into periods of two figures each, the number of periods thus formed will determine the number of figures in the root, and the left hand period, (81,)will contain the square of the left hand figure, (9,) of the root.

Hence, having pointed this number, (1296,) I will extract the root of the greatest square in the left hand period, and it will be the left hand figure of the root. I commence at the left hand also for the reason that if I have a remainder, I can reduce it to units of a lower denomination, and thus avoid fractions.

The greatest square number in 12 is 9, the root of which is 3. Subtracting this square from the left hand period, I have three for a remainder. Now as the next figure of the root affects the next two places in the square, I will annex to this remainder the next period of the square.

Now that I may obtain the next figure of the root, I will ascertain of what this remainder is composed. For that purpose I have taken the number 36 and squared it, representing the operation by appropriate signs. The tens multiplied by tens give that. (3×3.) The tens into the units give that. (3×6.) The units into the tens give that. (3×6.) The units into the units give that. (6×6.)

I notice that the square of a number consisting of tens and units, is composed of the square of the tens, (3^2,) plus twice the product of the tens by the units, ($2\times3\times6$,) plus the square of the units, (6×6.)

I have already subtracted the square of the tens, hence this remainder must be composed of twice the product of tens by the units, plus the square of the units : but mostly of twice the product of the tens by the units. Therefore if I divide this remainder by twice the tens, I shall obtain the units—or some number larger.

But twice the tens will give tens, and that multiplied by units will give tens, hence the units figure of this remainder will form no part of twice the tens by the units—therefore I will reject the units figure, and divide by twice the tens by the units.

Twice 3 is 6 ; and 6 is contained in 39 six times—but as this remainder is also composed of the square of the units, I will write the 6 at the right of the divisor, that it may be squared when I multiply. Now, multiplying and subtracting, I have no remainder.

The square of a decimal fraction always consists of full periods of two figures each : hence, commence at the right hand to point off in decimals, so that if there are not full periods we may annex naughts, and thus give it the *form* of a *complete square*.

Hence, to extract the Square Root:

Point the number into periods of two figures each, commencing at the decimal point and pointing each way. Find the greatest square number in the left hand period, and place its root as the left hand figure of the root. Subtract that square number from the left hand period, and to the remainder annex the next period for a dividend. Take twice the root already found for a divisor ; find the number of times the divisor is contained in the dividend, rejecting the right hand figure ; place the result as the next figure of the root, and also at the right of the divisor. Multiply the divisor thus augmented by the last figure of the root found, and subtract the product from the dividend. If there is no remainder or more periods, we have the entire root ; but if there *is* a remainder, annex the *next* period, and form a new divisor, by taking twice the root *now* found, and thus proceed until all the periods have been used.

EXTRACTION OF CUBE ROOT.

1	2	3	4	5	6	7	8	9
1	8	27	64	125	216	343	512	729

221	976
$200^2=4\,000\,000$	$900^2=729\,000\,000$
$20^2=\quad 4\,000$	$70^2=\quad 343\,000$
$1^2=\quad 1$	$6^2=\quad 216$

<table>
<tr>
<td>
3 6
3 6
———
$3\times3+3\times6$
$\quad 3\times6+6\times6$
———
3 6
———
$3\times3\times3+3\times3\times6$
$\quad +3\times3\times6+3\times6\times6$
$\quad +3\times3\times6+3\times6\times6$
$\quad +3\times6\times6+6\times6\times6$
———
$3^2 + 3^2\times6 + 3\times6^2 + 6^3$
</td>
<td>
13,824[24
8
———
12) 5 824
———
48
96
64
———
5824
———
0000
</td>
</tr>
</table>

I wish to find a means of extracting the cube root of a number—that is of finding a number, which, being used three times as a factor will produce the given number, (13,824.)

First, I have cubed each of the nine digits, and I find that no one of the cubes consists of more than three figures. But as there are more than three figures in this number, (13,824,) I know there must be more than one figure in the root.

I now wish to find in what places of the cube, the cube of each figure in the root will be found.

For that purpose I have taken the two numbers 221 and

976, and decomposed them into the parts 200, 20, and 1—and 900, 70, and 6, and have cubed those parts.

From our system of Notation, I know that the cube of the hundreds figure will occupy the seventh place from the right, and if the figure is of sufficient value, its cube may occupy the seventh, eighth, and ninth places from the right. That the cube of the tens figure will occupy the fourth place from the right, and, if the figure is of sufficient value, its cube will occupy the fourth, fifth, and sixth places from the right. That the cube of units will occupy the first place at the right, and if the figure is of sufficient value, its cube will occupy the first, second, and third places from the right. Now as the cube of tens cannot reach nearer to the right than the fourth place,—the cube of units will be found in the first three places at the right. And, as the cube of hundreds cannot reach nearer to the right than the seventh place, the cube of tens will be found in the next three places; the cube of hundreds in the next three places, &c.

Hence, if in the extraction of cube root, we point the number off into periods of three figures each, the number of periods thus formed will determine the number of figures in the root, and the left hand period will contain the cube of the left hand figure of the root.

Hence having pointed this number, (13824,) I will extract the root of the greatest cube in the left hand period and it will be the left hand figure of the root. I commence at the left hand, also for the reason, that if I have a remainder, I can reduce it to lower denominations and thus avoid fractions.

The greatest cube number in 13 is 8, the cube root of which is 2. Subtracting this cube number from the left hand period I have 5 for a remainder. Now as the next figure of the root affects the next three places in the cube, I will annex to this remainder the next period of the cube. Now, that I may obtain the next figure of the root, I will ascertain of what this remainder is composed. For that purpose I have taken the number 36 and cubed it,—representing the operation by appropriate signs. First, I must square it. The tens into the tens give that—$(3\times3.)$ The tens into the units, give that—$(3\times6.)$ The units into the tens give that—$(3\times6.)$ The units into the units give that—$(6\times6.)$ I now have the square of 36, which being multiplied by 36 will give its cube.

The tens into that line, $(3\times3+3\times6)$ give that—$(3\times3\times3+3\times3\times6.)$ The tens into that line, $(3\times6+6\times6,)$ give that—$(3\times3\times6+3\times6\times6.)$ The *units* into that line, $(3\times3+3\times6.)$ And the units into *this* line, $(3\times6+6\times6,)$ give that—$(3\times6\times6+6\times6\times6.)$

I notice that the cube of a number consisting of tens and units is composed of the cube of the tens, (3^3) plus three times the square of the tens by the units, $(3\times3^2\times6,)$ plus three times the tens by the square of the units, $(3\times3\times6^2,)$ plus the cube of the units.

I have already subtracted the cube of the tens—hence, this remainder must be composed of three times the square of the tens by the units,—three times the tens by the square of the units,—and the cube of the units. But a great part of it is made up of three times the square of the tens by the units. Hence, if I divide this remainder by three times the square of the tens, I shall obtain the units, —or some number larger.

But the square of tens gives hundreds,—and that multiplied by units gives hundreds,—hence the two right hand figures of the dividend can form no part of three times the square of the tens by the units,—therefore *reject* the two right hand figures, and divide the rest by three times the square of the tens.

The square of the tens (2) is 4, which multiplied by 3 gives 12 for a divisor. Twelve is contained in 58 four times,—place the 4 as the next figure of the root. Multiplying the divisor by this figure, (4,) I obtain three times the square of the tens by the units, (48,) agreeing with the first part of the remainder, which being hundreds, place it in hundreds place.

The next part of the remainder is composed of three times the tens into the square of units. Three times the tens (2,) gives 6, which multiplied by the square of units,

(16,) gives 96. Three times the tens gives tens, and the square of units being units, three times the tens into the square of units gives tens—hence, place it in tens place. The cube of units being units, place it in units place. Adding, and subtracting from the dividend, I have no remainder, hence, 24 is the exact root.

The cube of a decimal fraction always consists of full periods of three figures each—hence, commence at the left hand to point off in decimals, so that if there is not full periods we may annex naughts, and thus give it the *form* of a complete cube.

Hence, to extract the Cube Root:

Point the number into periods of three figures each, commencing at the decimal point and pointing each way. Find the greatest cube number in the left hand period, and place its root as the left hand figure of the root. Subtract that cube number from the left hand period, and to the remainder annex the next period for a dividend. Take three times the square of the root already found for a divisor. Find the number of times it is contained in the dividend, rejecting the two right hand figures. Place the result as the next figure of the root. Multiply the divisor by this figure, and place the product under that part of the dividend used. Then take three times the tens, (of the root,) into the square of the units, and place under the first product, one place farther to the right. Then take the cube of the units and place under the last product, one place *still farther* to the right. Add the three results together, and subtract their sum from the dividend. If we have no remainder or more periods, we have the entire root; but if there is a remainder, annex the *next* period, and form a new divisor by taking three times the square of the root *now* found, &c., and thus proceed until all the periods have been used.

DUODECMIALS.

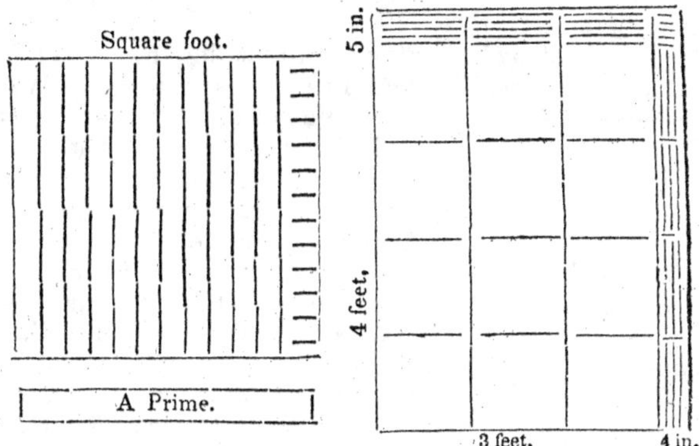

If a square foot be divided into twelve equal parts, each one foot long and one inch wide, the surfaces so obtained are called *primes.* If one of these primes be divided into twelve equal parts, each one inch long and one inch wide, the surfaces so obtained are *seconds.* If a second be divided into twelve equal parts we obtain *thirds*, &c.

Primes are designated by one stroke, seconds by two strokes, thirds by three strokes, &c.—thus, $3ft.\ 2'\ 7''\ 8'''$ is read three feet, two primes, seven seconds and eight thirds.

It is evident that twelve units of any one of these denominations will make one of the next higher. And as their values increase or decrease in a twelve-fold proportion, these numbers are called *Duodecimals*; and as *we* shall explain them, they are used in the measurement of surfaces, but they *may* be used in the measurement of solids also.

Suppose that I have a board four feet and five *inches* (not primes) long, and three feet and four inches wide, of which I wish to find the amount of surface.

If I give the four feet of length one foot of breadth, I shall obtain four square feet and five primes. This will extend across the board one foot, but as the whole board contains three and four-twelfths as much surface, I must multiply the

4ft.	5′ by	
3	4 twelfths.	
13	3′	
1	5′	8″
14	8′	8″

First, I will multiply by 3. Three times five primes are fifteen primes, equal to one foot and three primes. I will write the three primes and reserve the one foot.

Three times the four feet are twelve feet, and the one foot reserved being added gives thirteen feet, which I will write down.

Now I will multiply by the four-twelfths. Four times the five would be twenty *primes*, if the four was a whole number, but as it is twelfths, the product will be twelfths of primes or seconds—equal to one prime and eight seconds. I will write down the eight seconds one place towards the right, and reserve the one prime. Four times four would give sixteen feet if the four was a whole number, but as it is twelfths, the product will be twelfths of square feet, or primes, and the one prime being added gives seventeen primes—equal to one foot and five primes. I will write the five primes under the primes and the one foot under the feet. The first partial product represents the surface of the board which is three feet in width, and the second partial product represents the surface of the board which is four inches in width; hence, if I add them I shall obtain the whole surface of the board, 14*ft.* 8′ 8″. Hence, the rule:

Multiply one dimension by the number of units in the other, commencing with the left hand term of the multiplier. Carry one for every twelve, and place the product arising from the successive parts of the multiplier, one place farther towards the right. Then the sum of the partial products will be the entire product.

ALLIGATION.

If we mix 4 bushels of oats at 25 cents, with 5 bushels of corn at 40 cents, and 6 bushels of barley at 45 cents—what will one bushel of the mixture be worth?

4 bushels of oats	at 25 cents	is worth		$1 00
5 "	" corn " 40	"	" "	2 00
6 "	" barley " 45	"	" "	2 70
15 bushels of mixture		is worth		$5 70

Hence, one bushel of the mixture is worth one fifteenth part of $5 70, equal to 38 cents.

Hence, having the price and quantity of each of several simples given, to find the mean price of the mixture:

Multiply each price by the quantity and divide the sum of the products by the sum of the quantities.

This process is called *Alligation Medial.*

Hence, Alligation Medial is finding the mean price of a mixture, formed from simples whose quantities and prices are known.

When the price or rate of each simple is given, and we wish to find the quantity to be taken of each to make a mixture of a given rate or quality, the process is called *Alligation Alternate.*

21	18	3	4	12	18) 54 (3
	20	1	7	7	54
	23	2	2	4	
	24	3	(5)	15	
			18		

Suppose a grocer to have sugars—some at 18 cents, some at 20 cents, some at 23 cents, and some at 24 cents a pound, of which he wishes to make a mixture worth 21 cents a pound—how many pounds of each kind must he take?

If he takes one pound of the 18 cents sugar, puts it into the mixture and sells it for 21 cents, he will gain three cents.

If he takes one pound of the 20 cents sugar, puts it into the mixture and sells it for 21 cents, he will gain one cent.

If he takes one pound of the 23 cents sugar, puts it into the mixture and sells it for 21 cents, he will lose two cents.

If he takes one pound of the 24 cents sugar, puts it into the mixture and sells it for 21 cents, he will lose three cents.

If he should take four pounds of the 18 cents sugar, instead of taking but one, he would gain twelve cents.

If of the 20 cents sugar he should take seven pounds, he would gain seven cents.

If of the 23 cents sugar he should take two pounds, he would lose four cents.

We now find that he has gained in all nineteen cents, and that he has lost but four cents—hence, in order that the mixture shall be worth 21 cents a pound, he has yet 15 cents to lose—and he must lose it on the 24 cents sugar. If on one pound of the 24 cents sugar he lose three cents, it will take as many pounds, that he may lose 15 cents, as three is contained times in fifteen. It is contained five times. Hence, he may take five pounds of the sugar at 24 cents,—two pounds at 23 cents,—seven pounds at 20 cents,—and four pounds at 18 cents, to make a mixture worth 21 cents a pound.

Suppose we were required to make a mixture containing 54 pounds.

By adding the quantities now obtained, I find we have 18 pounds; just one-third of the quantity required—hence, I must multiply each quantity already found by 3.

Therefore, having the prices or rates of several simples given, to find the quantity to be taken of each to make a mixture of a given rate or quantity:

Write the prices in a column, with the smallest first, the next larger next, and so on,—and the mean rate at the left hand. Draw a horizontol line separating the prices less than the mean rate, from those that are greater. Find the difference between the prices and the mean rate, and place opposite the prices themselves. Multiply those differences, omitting one, by numbers taken at pleasure. Then from the sum of the products not on the same side of the horizontal line with the omitted difference, take the sum of the products *on* the same side,—divide the remainder by the omitted difference, and the quotient together with the quantities taken at pleasure, will be the proportionate quantity of each simple, to be taken.

If the whole quantity is limited :

Divide the quantity to which it is limited by the sum of the quantities found by the rule, and multiply each quantity thus found by the quotient—the products will be the quantities to be taken of each.

ARITHMETICAL PROGRESSION.

1st.	2d.	3d.	4th.	5th.
3	7	11	15	19
3	3+4	3+4+4	3+4+4+4	3+4+4+4+4

First term=3
Last term=19
Number of terms=5
C. d.=x

$$\begin{array}{r} 19 \\ 3 \\ \hline 4)\,16 \\ \hline 4 = \text{c. d.} \end{array}$$

First term=3
Last term=19
C. d.=4
Number of terms=x

$$\begin{array}{r} 19 \\ 3 \\ \hline 4)\,16 \\ \hline 4+1 = \text{No. terms.} \end{array}$$

First term=3
Last term=19
No. of terms=5
Sum of terms=x

$$\begin{array}{rrrrr} 3 & 7 & 11 & 15 & 19 \\ 19 & 15 & 11 & 7 & 3 \\ \hline 22 & 22 & 22 & 22 & 22 \\ & & & & 5 \\ \hline & & & 2) & 110 \\ \hline & & & \text{Sum} = & 55 \end{array}$$

A *series* is a succession of terms derived from each other according to some fixed and known law.

A series of numbers increasing by a common addition, or decreasing by a common subtraction, is called an *Arithmetical Progression*.

The common number added or subtracted is called the *common difference.*

The first and last terms are called, *extremes.*

The first term (3) is a term given. The second term (7) is formed by adding the common difference to the first term. (3+4). The third term (11) is formed by adding the common difference to the second term (3+4+4,) —but as the second (7) has the common difference occurring once, the third term will have the common difference occurring twice.

The fourth term is found by adding the common difference to the third term—but as the third term has the common difference occurring twice, the fourth term will have the common difference occurring three times.

The fifth term is formed by adding the common difference to the fourth term—but as the fourth term has the common difference occurring three times, the fifth term will have the common difference occurring four times.

I notice the common difference occurs *four* times in the *fifth* term, *three* times in the *fourth* term, *twice* in the *third* term, &c. And in general, the common difference occurs in any term a number of times indicated by the number of terms less one. Hence, to find any term:

Multiply the common difference by the number of the term less one, and to the product add the first term.

Suppose we have given, the first term, the last term, and the number of terms, to find the common difference.

The last term (19) is made up of the common difference (4) taken a number of times indicated by the number of terms less one, added to the first term. Hence, if I subtract the first term from the last, the remainder (16,) will be composed of the common difference, taken a number of times indicated by the number of terms less one. Therefore if I divide this remainder (16) by the number of terms less one, (4,) I shall have the common difference. Hence, having the first term, the last term, and the number of terms given, to find the common difference:

Divide the difference of the extremes by the number of terms less one.

Suppose that I have given, the first term, the last term, and the common difference, to find the number of terms.

The last term (19) is made up of the common difference (4) taken a number of times indicated by the number of terms less one, added to the first term. Hence, if I subtract the first term from the last, the remainder (16) will be composed of the common difference taken a number of times indicated by the number of terms less one. Therefore, if I divide this remainder (16) by the common difference, (4,) I shall have the number of terms less one—and adding 1 I obtain the number of terms. Hence, having the first term, the last term, and the common difference given, to find the number of terms :

Divide the difference of the extremes by the common difference, and add one to the quotient.

Suppose that I have given, the first term, the last term, and the number of terms, to find the sum of the series.

If the first and last terms be added together we shall obtain a certain amount. (22.) If we take the second term, counting from the left, which is *greater* than the first term by the common difference, and add it to the second term, counting from the *right*, which is *less* than the last by the common difference, we shall obtain the same amount. (22.)

Again—if we take the third term, counting from the left, which is greater than the *second* by the common difference, and add it to the third, counting from the right, which is *less* than the second by the common difference, we again obtain the *same amount:* (22.)

And, in general, the sum of the extremes is equal to the sum of any two terms equally distant from the extremes. Hence, if I invert the series and add it to itself, I shall obtain a set of *like numbers*—the sum of which may be obtained by multiplying one of them (22) by the number of them. (5.)

But this set of like numbers was formed by adding the original series to itself—hence, it must be twice the original series—therefore its *sum* (110) is twice the sum of the original series,—and hence, divide this sum (110) by 2, and we have the sum of the original series.

Hence, having given, the first term, the last term, and the number of terms, to find the sum of the series:

Multiply the sum of the extremes by the number of terms, and divide the product by 2,—or multiply the sum of the extremes by one-half the number of terms, or multiply one-half the sum of the extremes by the number of terms.

GEOMETRICAL PROGRESSION.

1st.	2d.	3d.	4th.	5th.
2	8	32	128	512
2	2×4	$2\times4\times4$	$2\times4\times4\times4$	$2\times4\times4\times4\times4$
2	2×4	2×4^2	2×4^3	2×4^4

First term=2
Last term=512 2) 512
No. of terms=5
Ratio=x $\sqrt[4]{256}=4$ Rat.

$4 : x :: x : 16$
$x^2=64$
$x=8$

First term=2 2 8 32 128 512
Last term=512 8 32 128 512 2048
Ratio=4 2
Sum=x

3) 2046
682 Sum.

A series of numbers increasing by a common multiplier or decreasing by a common divisor, is called a *Geometrical Progression.*

The common multiplier or divisor is called the *Ratio.*

The first term is a term given. The second term is found by multiplying the *first* term by the *ratio.* The third term is found by multiplying the second term by the ratio; but as the ratio occurs in the second term once as a factor, it will occur in the *third* term *twice* as a factor, or raised to the second power. The fourth term is found by multiplying the third term by the ratio; but as the ratio occurs in the third term twice as a factor, it will occur in the *fourth* term *three* times as a factor, or raised to the third power. The fifth term is found by multiplying the fourth term by the ratio; but as the ratio occurs in the fourth term three times as a factor, it will occur in the *fifth* term *four* times as a factor, or raised to the fourth power.

We may notice that the ratio occurs raised to the *fourth* power in the *fifth* term, to the *third* power in the *fourth* term, to the *second* power in the *third* term, &c. And in general, the ratio occurs in any term, raised to a power indicated by the number of the term less one.

Hence, to find any term:

Raise the ratio to a power indicated by the number of the term less one, and multiply *that power* by the *first term.*

Suppose that we have given the first term, the last term, and the number of terms, to find the *ratio.*

The last term is made up of that power of the ratio indicated by the number of terms less one, multiplied by the first term: hence, if I divide the last term by the first, I shall have for a quotient that power of the ratio indicated by the number of terms less one. Therefore if I take that root of the quotient indicated by the number of terms less one, I shall have the ratio.

Hence, having given the first term, the last term and the number of terms, to find the ratio:

Divide one extreme by the other, and take that root of the quotient indicated by the number of terms less one.

Suppose I wish to find the mean proportional between two numbers, as 4 and 16. As both means of the proportion are wanting, we have $4 : x :: x : 16$. But as the product of the means is equal to the product of the extremes, we have the equality, $x^2=64$. Now if $x^2=64$, the root of x^2 must be equal to the root of 64, (or 8.) Hence, 8 is the mean proportional sought.

Therefore the rule to find a mean proportional between two numbers :

Multiply them together and take the square root of their product.

Suppose we have given the first term, the last term and the ratio, to find the sum of the series.

If I multiply the original series by *its own ratio*, I shall obtain a series which will be as many times the original series as there are units in the ratio. Now if I subtract the first series from this new series, I shall obtain as many times the original series as there are units in the ratio less one. Subtracting, the terms cancel, except the first and last. Subtracting the first (2) from the last (2048,) I obtain 2046 ; which is as many times the first series as there are units in the ratio less one. Hence, *divide* by the number of units in the ratio less one, and we have the sum of the series.

Hence, the rule :

To find the sum of the series, multiply the last term by the ratio ; from that product subtract the first term, and divide the remainder by the ratio less one.

ANNUITIES AT SIMPLE INTEREST.

An *Annuity* is a sum to be paid at stated periods.

When an annuity remains unpaid for some number of years, it is said to be *in arrears*, or to be *forborne*.

I wish to show that the book account of an annuity in arrears, at simple interest, constitutes an *Arithmetical Series*.

Suppose the annuity to be $50, to be paid annually, but that it has remained unpaid for some years, simple interest being reckoned, at 6 per cent.

50		At the expiration of the first year, there
50	3	will be charged $50 for annuity. At the end
50	6	of the second year there will be charged for
50	9	annuity $50, and $3 for interest on the first
&c.	*&c.*	$50 for the last year. At the end of the third

year there will be charged for annuity $50, and $6 for interest on the last two fifties for the last year. At the end of the fourth year there will be charged for annuity $50, and $9 for interest on the last thee fifties for the last year.

By examining these charges we shall find that they form an Arithmetical Progression, of which the annuity is the first term, the interest on the annuity for one year is the common difference, and the number of years is the number of terms.

ANNUITIES AT COMPOUND INTEREST.

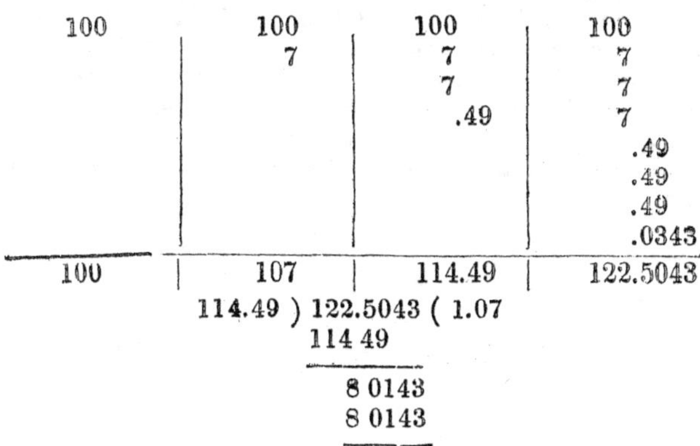

100	100	100	100
	7	7	7
		7	7
		.49	7
			.49
			.49
			.49
			.0343
100	107	114.49	122.5043

114.49) 122.5043 (1.07
114 49

8 0143
8 0143

I wish to show that the book account of an annuity in arrears at compound interest constitutes a *Geometrical Series.*

Suppose the annuity to be $100, to be paid annually, but that it has remained unpaid for some years, and compound interest to be reckoned at seven per cent.

At the expiration of the first year, there will be charged $100, for annuity. At the end of the second year, there will be charged for annuity $100 and $7 for interest on the first $100 for the last year. At the end of the third year, there will be charged for annuity $100 and $7 for interest on the first $100 charged, for the last year, and $7 for interest on the second $100 for the last year, and .49 for interest on the $7 charged for the last year. At the end of the fourth year there will be charged $100 for annuity, $7 for interest on the first $100 for the last year,—$7 for interest on the second $100 for the last year,—$7 for interest on the third $100 for the last year—.49 for interest on the first $7 of interest for the last year—.49 for interest on the second $7 for the last year—.49 for interest on the third $7 for the last year—and .0343 for interest on the .49 for the last year. Now if any one of these annual charges, be divided by the one immediately preceding it, the quotient will be the amount of one dollar for one year at the given rate per cent.

Hence, the book account of an annuity in arrears at compound interest *does* form a geometrical series of which the annuity is the first term,—the amount of one dollar for one year is the ratio,—and the number of years is the number of terms.

THEORIES OF ALGEBRA.

ADDITION.

$4a$	$+6c$	$+7ab$	$-7ab$	$+12ab$	$-12ab$
$5b$	$-4d$	$+5ab$	$-5ab$	$-7ab$	$+7ab$
$4a+5b.$	$6c-4d.$	$+12ab.$	$-12ab.$	$+5ab.$	$-5ab.$

To $4a$ I wish to add $5b$. As the quantities are dissimilar I can only add them by writing the proper sign between them.

To plus 6c I wish to add minus $4d$. The $4d$ being a subtractive quantity, if I annex it to the $6c$ it will evidenly diminish its value, and the result will be $6c-4d$.

To plus $7ab$ I wish to add plus $5ab$. As the quantities are similar, and both plus, it is evident that I shall have a plus amount equal to the sum of the two, or $+12ab$.

To minus $6ab$ I wish to add minus $5ab$. As the quantties are similar, and both minus, it is evident that I shall have a minus amount equal to the sum of the two, or $-12ab$.

To plus $12ab$ I wish to add minus $7ab$. The minus sign before the $7ab$ shows it to be a subtractive quantity ; hence, if I unite it to the $12ab$ it will *take from* it, and I will evidently have $+5ab$.

To minus $12ab$ I wish to add plus $7ab$. The minus sign before the $12ab$ shows the $12ab$ to be a subtractive quantity. Hence, if I unite it to the $7ab$, it will take $12ab$ from it. But I cannot take the whole of $12ab$ from $7ab$. The $12ab$ is composed of $7ab$ and $5ab$—I can take $7ab$ from the $7ab$ and there will remain $5ab$ yet to be subtracted —to *show* that it is to be subtracted, write the minus sign before it. Now as these are all the cases that can occur, we have the rule :

When the quantities are dissimilar, write them after each other with their original signs.

When the quantities are similar and their signs alike—add their coeficients, annex the literal part, and prefix the sign of the quantities themselves.

When the quantities are similar and the signs *unlike*, take the less coeficient from the greater, to the remainder annex the literal part, and prefix the sign of the greater coeficient.

SUBTRACTION.

$5a$
$4b-3c$
$5a-4b+3c$

Let $4b$ represent the sum of all the *additive terms* in the subtrahend, and $3c$ the sum of all the *subtractive terms*. Then from $5a$ I wish to take $4b$ minus $3c$. Not the whole of $4b$, but $4b$ diminished by the number of units in $3c$.

First I *will* take from $5a$ the *whole* of $4b$, which gives $5a-4b$. Now as the subtrahend has been too large by $3c$, the remainder is too small by $3c$—hence, to make it what it should be, I must add $3c$—which gives $5a-4b+3c$. I notice that the plus sign of the subtrahend has been changed to minus, and the minus sign has been changed to plus, and then it has been added to the minuend. Hence, for subtraction:

Change all the signs in the subtrahend, and then proceed as in Addition.

MULTIPLICATION.

$7a^4b^3c^4$
$5a^3b^3c^5$
$35a^7b^6c^9$

I wish to multiply one monomial by another—for instance, $7a^4b^3c^4$ by $5a^3b^3c^5$. Multiplication consists in uniting with the multiplicand such factors as occur in the multiplier. If with the factor 7 I combine the number 5 as a factor, I shall have for the product 35. As a occurs four times as a factor in the multiplicand, and three times in the multiplier, if I unite them as *factors*, it will occur seven times in the product. As b occurs three times as a factor in the multiplicand and three times in the multiplier, if I unite them as factors, it will occur six times in the product. As c occurs four times as a factor in the multiplicand and five times in the multiplier, if I unite them as factors it will occur nine times as a factor in the product. Now each letter will occur as many times as a factor in the product, as it does in both multiplicand and multiplier taken together. But as the exponents show the number of times each letter occurs as a factor—to find the exponent of any literal factor in the product, add the exponents of that letter in multiplier and multiplicand. Hence, to multiply one monomial by another:

Multiply the coefficients together for a new coefficient, write after that all the literal factors, affecting each with an exponent equal to the sum of the exponents in both the factors.

THE SIGNS IN MULTIPLICATION.

$+5b$	$-b$,	$-b$,	$-b$,	$-b$,	$-b$,	$-b$,	$-b$
$+3a$	3	2	1	0,	-1,	-2,	-3
$+15ab$	$-3b$,	$-2b$,	$-b$,	0,	$+b$,	$+2b$,	$+3b$,

I wish to find what sign to give the product in multiplication.

Plus $5b$ multiplied by plus $3a$ will evidently give a positive product. ($15ab$.)

$-b$ multiplied by 3 will evidently give three times as much minus, or $-3b$.

$-b$ multiplied by 2 will evidently give three times as much minus, or $-2b$.

$-b$ taken once will evidently give $-b$.

$-b$ multiplied by naught will give naught.

Now while the multiplicands have each been of the same value, and the multipliers have been diminishing by a comdifference of 1, the products have been increasing by a common difference of b.

And as the multiplicands that yet follow, have the same value, and the multipliers *continue* to diminish by a common difference of 1, I conclude that the products must *continue to increase* by a common difference of b—but if they do, the next one will be plus b, the next plus $2b$, the next plus $3b$, &c. Hence minus b multiplied by minus 3 must give plus $3b$.

Therefore we conclude that when the signs of the factors are *alike*, the product will be *plus*, and when the signs of the factors are *unlike*, the sign of the product will be *minus*.

MUTIPLICATION OF POLYNOMIALS.

$$\begin{array}{ll} a—b & \quad a—b \\ c—d & \quad\quad d \\ \hline ac—bc—ad\times bd & \quad ab—bd \end{array}$$

I wish to multiply one polynomial by another.

Let a represent the sum of all the additive terms in the multiplicand, and b the sum of the subtractive terms. Let c represent the sum of all the additive terms in the multiplier, and d the sum of all the subtractive terms. Then I wish to multiply $a—b$ by $c—d$—not by the *whole* of c, but by c diminished by the number of units in d.

First, I will multiply by the whole of c. c times a gives ac—but before I multiply by c the *a should be diminished* by b—hence, in multiplying the *whole* of a, I have multiplied a b which I *should not* have multiplied, and hence the product is too large by a times b. Hence, I must subtract c times b or ab from it, which gives $ac—bc$. Therefore c times $a—b$ is $ac—bc$.

But now I have multiplied by a quantity too large by the units in d, hence this product ($ac—bc$) is too large by d times $a—b$—therefore I must obtain d times $a—b$, and subtract from this product. d times $a—b$ gives $ad—bd$, which being subtracted from $ac—dc$ gives $ac—bc—ad\times bd$. I notice that each term in the multiplicand has been multiplied by each term of the multiplier, hence the rule :

Multiply each term in the multiplicand by each term in the multiplier,—affecting each partial product with the plus sign when the signs of the factors are alike, and with the minus sign when the signs of the factors are unlike.

DIVISION OF MONOMIALS.

$$7a^2b^3)35a^4b^8c$$
$$5a^2b^5c$$

I wish to divide $35a^4b^8c$ by $7a^2b^3$. Division consists in having two terms given to find a third term, which being multiplied by the first will produce the second. The first is called the divisor, the second the dividend, and the third the quotient. Hence, the dividend may always be considered as a *product* of the other two terms.

From multiplication of monomials, we know that the coefficient of the dividend is made up by multiplying the coefficient of the divisor by the coefficient of the quotient—hence, if I divide the coefficient of the dividend by the coefficient of the divisor, I shall obtain the coefficient of the quotient.

The exponont of any letter, as a in the dividend, is made up by adding the exponent of that letter in the divisor, to the exponent of the same letter in the quotient—hence, if I subtract the exponent of a in the divisor from the exponent of the same letter in the dividend, I shall have the exponent of that letter in the quotient. And in the same manner with b,—if I subtract the exponent of b in the divisor from the exponent of b in the dividend, I shall have the exponent for b in the quotient. The c must occur in the quotient, for if it does not, to multiply the divisor and quotient together will not give the dividend. Hence, the rule for the division of monomials:

Divide the coefficient of the dividend by the coefficient of the divisor for a new coefficient—write after that all the literal factors of the dividend, affecting each with an exponent equal to the excess of its exponent in the dividend over that in the divisor.

Remark.—Division of monomials cannot be performed,

when the divisor contains a factor which is not found in the dividend—and when a factor in the divisor has a larger exponent than the exponent of the same letter in the dividend—and when the coefficient of the dividend will not contain the coefficient of the divisor.

DIVISION OF POLYNOMIALS,

$$3a^5b+2a^3b^3-3ab^6 \qquad 26a^2b^2+10a^4-48a^3b+24ab^3 \;\big|\; 4ab-5a^2+3b^2$$
$$2a^4b^2+3a^2b^4+2a^2b^2 \qquad -6a^2b^2+10a^4-8a^3b \;\big|\; -2a^2+8ab$$
$$6a^9b^3+\text{\&c.} \qquad 32a^2b^2-40a^3b+24ab^3$$
$$32a^2b^2-40a^3b+24ab^3$$

Second form.
$$10a^4-48a^3b+26a^2b^2+24ab^3 \;\big|\; -5a^2+4ab+3b^2$$
$$10a^4-8a^3b-6a^2b^2 \;\big|\; -2a^2+8ab$$
$$-40a^3b+32a^2b^2+24ab^3$$
$$-40a^3b+32a^2b^2+24ab^3$$

In the product arising from the multiplication of one polynomial by another, there will always be some irreducible terms—and one of those terms will be the one formed by multiplying that term in the multiplicand, containing the largest exponent of a letter, by the term in the multiplier containing the largest exponent of the *same* letter. For, we shall thus obtain a term which will contain a greater exponent of that letter, than the exponent of that letter in any other term of the product; hence, it will not be like any other term, and hence it will be irreducible.

This is equally true for all products of polynomials. Hence, if we examine any product and select the term which contains the largest exponent of a letter in that product, we know that such term is made up by multiplying the term in one of the factors containing the largest exponent of that letter by the term in the other factor containing the largest exponent of the same letter.

All dividends are products: hence, if I trace along this dividend and find the term containing the largest exponent of some letter, (as a,) I know that it is made up by multiplying the term in the divisor containing the largest exponent of a, by the term containing the largest exponent of a in the quotient. Hence, if I divide that term in the dividend containing the largest exponent of a, by the term in the divisor containing the largest exponent of a, I shall obtain a term for the quotient containing a larger exponent of that letter than any other term in the quotient. Dividing $10a^4$ by $-5a^2$ gives $-2a^2$ for a term in the quotient. Now, as the dividend is made up by multiplying the divisor by the quotient, if I multiply the divisor by the *part* of the quotient found, I shall obtain a part of the dividend; and subtracting that product from the whole divideud, I obtain that part of the dividend which is made up by multiplying the divisor by that part of the quotient not yet found: hence, this *remainder* is a *product*.

But from what was first shown, that term in it which contains the largest exponent of a, is made up by multiplying the term containing the largest exponent of a in the divisor, by that term containing the largest exponent of a in the part of the quotient not yet found. Hence, if I divide that term by the term in the divisor containing the largest exponent of a, I shall obtain that term containing the highest exponent of a in the part of the quotient not yet found.

Dividing $-40a^3b$ by $-5a^2$ gives $8ab$. Multiplying the *divisor* by the $8ab$, and subtracting, we have no remainder. Hence, $-2a^2+8ab$ is the entire quotient.

It has been necessary to compare the terms in the divisor containing the largest exponent of some letter with the term in the dividend containing the largest exponent of the same letter; and then the same term in the divisor with the term in the dividend containing the next smaller exponent, &c. Hence the rule for division of polynomials:

Arrange both divisor and dividend in reference to some letter, placing the term in each containing the largest exponent of some letter, at the left, the term containing the next smaller exponent next, &c.: thus the exponents of that letter will diminish from the left to the right : then divide the left hand term of the dividend by the left hand term of the divisor, for the left hand term of the quotient: multiply the divisor by that term of the quotient and subtract the product from the dividend : if there is a remainder, treat it as a new dividend, with which proceed as before.

NOTE.—When you subtract, be careful to keep the *highest* power of the *same* letter at the *left*, that you had at the left at first.

From Comstock's Phonetic Magazine.

ΔE FONΛK GURL OV MΛCΛGAN.

[This Poem is founded on the following fact, communicated in a letter from the Hon. Ira Mayhew of Michigan, to Dr. A. Comstock of Philadelphia:

"I took up the TREATISE ON PHONOLOGY, and I was unable to lay it down until I had completed its perusal. I placed it in the hands of my children, and judge what was my surprise to hear a daughter, not yet six years old, read the first chapter of Genesis fluently, in less than one hour from the time she first saw a Phonetic character!"]

A dɷtur hɛm prɸd kɩŋz wʊd ɷn,
  Adɷrnz ðɛ lek-gurt Stet.
Ər sumurz sɩx had ɷ'ur hur cɷn,
  Cx ɷp. ðɛ Fɷnɩon'z get.
Yɩs; ɷpt ðɛ Tɛmpl ov ɸr tuŋ,
  Ɩn wun cɷrt ɸr ɩndxd,
And lurnd ðɛ lɛturz qɩč ðɛ yuŋ,
  Bi Fonɩk skɩl, me rxd!

Δɛn stand rxbɯkt bi čildhʊd'z fem,
  Yx mɛn hʊ ɛrur hxd;
And lɛt an ɩnfant's progrɛs cem
  Yɯr livz ov pɷltrɩ dxd!
Se not, az yx hav sɛd tʊ mx,
  Δat "God haz kurst ɸr spxč,
And Bebɛlizd ɩt hɷplɛslɩ!"
  Nɷ mɷr ðɩs fɷlshʊd txč.

Se not, "Fɷnɛtɩks ar abstrɯs,—
  Abuv ðɛ pxpl'z mind!"—
A lɩtl gurl haz lurnd ðɛ yɯs
  Ov lɛturz trɯθ dxsind;
A lɩtl gurl haz lurnd tʊ rxd
  Bi sɩstɛm nxt and plen—
Cʊd not ðɩs sɩstɛm sɯpursxd
  Δɷz ov loŋ yxrz ov pen?

Akɯz not Hɛvn'z Gospɛl brit,
  Ov ɷlturɩŋ ɸr spxč!
Fɷr *ɩnɷsɛns* ɩz Gospɛl-lit;
  Ɩts akcunz vurčɯ txč!
Cal not ðɩs Alfabɛt bx ɷnd
  Bi Frxdum'z Stets rxnɸnd!
Δɛ Gospɛl, ɩn ɩt, trumpɩt-tɷnd,
  Gɩvz nɷ unsurtɩn sɸnd!

UNIVERSITY OF MICHIGAN
3 9015 04872 4242

REMARKS.

At the time of recitation, it is designed that the example given at the commencement of the explanation, should be copied upon the black-board, and that the student should be required to refer to it with the pointing stick. Let the explanation be committed to memory VERBATIM,—as the closer the student follows it, the better will be his success, and the more accurate and effectual habits of thought will he have acquired.

As the example for proportion was omitted at its proper place, we will insert it here.

yds.	$	yds.	$
2	6	9	
2	6	9	27
2 :	6 : :	9 :	27

2	6	$2 \times 27 = 9 \times 6$
2	6	$2 \times x = 9 \times 6$
27	9	$2 \times 27 = x \times 6$
54 = 54		

It will be noticed that the pronoun I is used in most cases in the explanations. This has been thought best, as it places the student in the place of an inquirer or investigator, and represents him as plodding his own way through the darkness of ignorance, and, if possible, into the light of simple truth.

We trust that the work will be heartily welcomed by that host of teachers who have so urgently requested its publication—and the hope that it shall be found USEFUL in the place designed it, has given a sweetness to the extra toil that it has brought upon

THE AUTHOR.

www.ingramcontent.com/pod-product-compliance
Lightning Source LLC
LaVergne TN
LVHW021428110826
845150LV00007B/2140